# Nanotechnology

# Nanotechnology
## Research and Perspectives

Papers from the First Foresight Conference
on Nanotechnology

edited by BC Crandall and James Lewis

The MIT Press
Cambridge, Massachusetts
London, England

This book is set in Sabon and was printed and bound in the United States of America.

Library of Congress Cataloging-in-Publication Data

Foresight Conference on Nanotechnology (1st : 1989 : Palo Alto, Calif.)
  Nanotechnology : research and perspectives : papers from the First Foresight Conference on Nanotechnology / edited by BC Crandall and James Lewis.
  p.    cm.
  Includes bibliographical references and index.
  ISBN 0-262-03195-7
  1. Nanotechnology—Congresses.  I. Crandall, BC  II. Title.
T174.7.F67   1989
620.4—dc20                                                    92-17183
                                                                    CIP

# Contents

# Preface

The papers collected in this volume evolved from presentations made at the First Foresight Conference on Nanotechnology held in Palo Alto, California, in October, 1989. This meeting, the first to directly address the technical challenges and social consequences of molecular engineering, brought together individuals from disparate areas of technical and societal research to discuss and debate the emerging field of nanotechnology. The material presented at that gathering has been transcribed, edited, and, in several cases, substantially revised and rewritten by the participants. The present volume is the culmination of those original talks and subsequent discussion and reflection.

From one point of view, nanotechnology is simply a descriptive term for a particular state of our species' control of materiality. At some point (seemingly in the next few decades) we will be able to create complex systems and structures whose individual components are best measured in billionths of a meter. This capacity arises as part of the natural evolution of homo faber's deep seated urge to manipulate matter. But the potential power of nanotechnology offers not only a quantum leap in our capacity to mold matter to our will but also a qualitative reorganization of our interactions with the world. If indeed it becomes possible to inexpensively create cubic-centimeter computers processing $10^{14}$ million instructions per second (MIPS) and self-reproducing, submicron industrial robots, what material reorganization won't be possible?

In a certain sense, all "machines" perform "magic." Both words (which share a linguistic root) express a capacity to manifest desire in actuality. Our current technological magic is very effective: I can easily whisper goodnight to a loved one on the far side of the planet. Machines are the means by which we change reality to more closely approximate our imagination of how things should be, and it seems that the molecular machines of nanotechnology will increase, by orders of magnitude, our individual and collective capacity to transform desires into material reality. Our world has changed dramatically under the auspices of modern technology. Molecular nanotechnology portends even more thorough transformation.

The workings of nanotechnology are new in that individual atoms will serve as building blocks. Hitherto all technology has utilized huge numbers of atoms. We are distinct from all previous generations in that we have *seen* our atoms—with scanning tunneling and atomic force microscopes. But more than simply admiring their regular beauty, we have begun to build minute structures. Each atom is a single brick; their electrons are the mortar. Atoms, the ultimate in material modularity, provide the stuff of this new technology.

This volume presents material for technical as well as cultural and ethical consideration. The potential for nanotechnology seems truly astronomical—as do the potential problems arising from its mismanagement or misuse. If we can find not only the intelligence but the wisdom to implement a globally beneficial nanotechnology, the future looks promising indeed.

As the editor of this volume, it has been my pleasure to work with the conference participants as their efforts developed into the chapters you are about to read. Their enthusiasm and concern for the emerging field of molecular nanotechnology was a continual source of inspiration. James Lewis, who transcribed the original presentations, provided invaluable technical and editorial assistance. Graham Walker graciously assisted in the creation of the index. Terry Ehling, of the MIT Press, whose genuine interest in nanotechnology and irrepressible encouragement brought many a day to light, provided the most, and least expressible, sustenance. Thank you all.

BC Crandall
Mill Valley, California

*The First Foresight Conference on Nanotechnology, chaired by K. Eric Drexler, was cosponsored by the Foresight Institute and the Global Business Network and was hosted by the Stanford University Department of Computer Science. Financial support was generously provided by the E. I. du Pont de Nemours & Co., Inc.*

*The Global Business Network, an international think tank and consulting firm, was founded to bring together information and perspectives relevant to the formulation of corporate strategy. The Foresight Institute was founded to explore and communicate the technological, cultural, and policy implications of molecular nanotechnology. The institute continues to hold conferences, teach, and publish material related to this emerging field. For ongoing information about nanotechnology, contact the institute directly.*

Foresight Institute
P.O. Box 61058
Palo Alto, CA 94306
Tel: (415) 324-2490
E-mail: foresight@cup.portal.com

# Nanotechnology

# 1

# Overview and Introduction

K. Eric Drexler

## Historical Perspective on Nanotechnology

The efforts to get better control of the structure of matter are ancient. For centuries people have rearranged patterns of atoms by heating, mixing, and pounding. In this century many of our most impressive technologies have continued to follow this pattern. Electronics have never been made using a careful molecular construction process but rather by heating, melting, and chipping. Even the finest integrated circuits exploit diffusion processes, with their inherent atomic-level randomness. The patterning of such surfaces is therefore statistical in nature, involving processes such as etching and vapor deposition.

Approaching the current limits of what has been achieved using bulk technologies of this sort—although falling rather short of what has been achieved since this figure was made—are the features shown in figure 1.1. These rather blurry lines say "NRL (for Naval Research Laboratories) MOLECULAR DEVICES." This is a pattern written in the surface of a salt crystal using a tightly focused electron beam. At Cornell, they have succeeded in writing rather finer lines, again by miniaturizing bulk technologies.

Such developments, however, are not the focus of this book. Etymologically, this *is* a form of nanotechnology: it involves fabrication at a nanometer scale, as shown by the 18 nm scale bar. However, the work presented here is on an emerging field in which a cubic nanometer is not a small element of resolution in an attempt to write patterns on a surface but a volume that contains many distinct atoms (figure 1.2). Each such

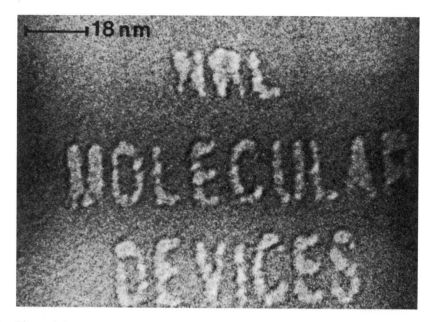

**Figure 1.1**
Text etched into a salt crystal with an electron beam. (Courtesy Naval Research Laboratories.)

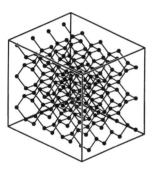

**Figure 1.2**
One cubic nanometer of diamond, containing 176 atoms. (A cube one micron on a side would contain 176 billion atoms and could contain a system of substantial complexity.)

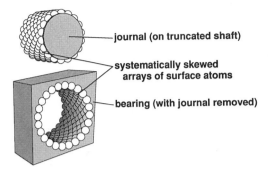

journal (on truncated shaft)

systematically skewed
arrays of surface atoms

bearing (with journal removed)

**Figure 1.3**
Steric repulsion cylinder-and-sleeve bearing.

element can serve as a distinct component in a complex structure—if one has sufficiently capable tools.

Our interest is also in devices that cannot be made with our present manufacturing technologies. Figure 1.3 illustrates an example from one such class of devices. The two surfaces consist of ordered arrays of covalently bound atoms. The shaft, when placed inside the tight-fitting sleeve, experiences a repulsive interaction with an energy of many electron volts. To build such a device and make it part of a mechanical system is beyond any conceivable lithographic technology and far beyond the state of the art of present-day chemistry. It will demand a molecular manufacturing technology based on direct, positional control of chemical reactions.

The field of "molecular nanotechnology" can be viewed as an elaboration of chemistry, in the same sense that modern computer technology can be viewed as an elaboration of electromagnetism and solid state physics: the most fundamental principles are unchanged, but new issues and capabilities emerge at higher levels of organization.

Such fabrication capabilities are still distant. The most relevant research today is in chemistry, biochemistry, and micropositioning technologies (such as those based on the scanning tunneling and atomic force microscopes). Some possible ways in which molecular nanotechnology could evolve from present-day capabilities are outlined in figure 1.4. The arrows represent (on the left) the development in chemistry and biochemistry of ever larger controlled molecular assemblages, and (on the

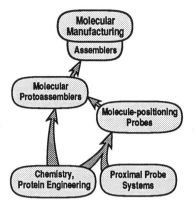

**Figure 1.4**
Molecular systems engineering and nanotechnology.

right) lines of advance toward integration of differing technologies. These steps lead toward molecular nanotechnology and manufacturing through molecular systems engineering.

## Molecular Systems Engineering

The five chapters in this section address aspects of the current technology base that are applicable to the emerging discipline of molecular systems engineering. One approach to building molecular systems is to exploit technologies that permit direct manipulation of individual atoms and molecules by macroscopic devices. This approach is the topic of chapter 2, by John Foster. Part of this work, imaging a graphite surface to which a single molecule has been covalently pinned, has graced the cover of *Nature*.[1] He discusses the use of scanning probes of solid matter to characterize surfaces in atomic detail and to perform structural modifications of those surfaces.

The most versatile material that biological evolution has discovered is protein. In chapter 3, Tracy Handel presents work on protein engineering. Protein engineering is a recent technology. As it is usually practiced, it involves altering natural proteins to make molecules with altered or improved characteristics. A related topic, not covered in this text yet quite relevant, is the use of the animal immune system to generate proteins (in this case, antibody molecules) with novel binding capabilities

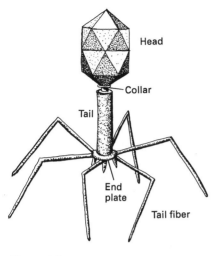

**Figure 1.5**
T4 bacteriophage assembly.

and even enzymatic activity.[2] In contrast to these approaches based upon proteins existing in nature, Tracy Handel discusses engineering "from scratch" nanometer-scale objects made of protein.

A crucial component of nanotechnology will be the design of molecules that will assemble into supramolecular structures. In chapter 4, Michael Ward describes the design and synthesis of self-assembling molecular structures. Ward works with synthetic organic molecules. Biological systems exploit self-assembly of larger molecules, proteins, and nucleic acids, to make structures like the T4 bacteriophage (figure 1.5); a glance at this structure certainly suggests the feasibility of self-assembling molecular machines. The principles of molecular self-assembly are fundamental to extending solution-based chemical synthesis chemistry from making small molecules to constructing large-scale molecular systems.

In chapter 5, Hiroyuki Sasabe provides a perspective on the growing research activities in Japan that are leading in the direction of molecular nanotechnology.

In chapter 6, I discuss strategies for engineering molecular systems, focusing on steps that can be taken from our present technology base in chemistry, biotechnology, and scanning-probe microscopy, and then take a brisk walking tour through some longer-term possibilities along the road to molecular manufacturing and molecular nanotechnology.

## Related Technologies

Chapters 7 through 9 focus on issues raised by nanoscale computation. Robert Birge describes what molecular systems engineering is accomplishing in molecular electronics today. Federico Capasso discusses developments in quantum electronics and semiconductor systems, including the potential for quantum electronics in structures that can not be fabricated using conventional technologies. Norman Margolus describes what has been learned about theoretical limits to computation, limits that molecular systems may help us to approach more closely. These topics are relevant to understanding the era of "post-micro-lithography electronics"—an era that lurks somewhere beyond the present business-planning horizon (at least on this side of the Pacific)—which will make a tremendous impact on the multibillion dollar industry of computational machines.

In chapter 10, Joseph Mallon presents a view of molecular nanotechnology from a micromachinist's perspective, clarifying the similarities and considerable differences between these approaches to the construction of small-scale devices.

Tracy Handel, John Foster, Ralph Merkle, and Michael Ward join me in a panel discussion (chapter 11) of the problems facing implementation of molecular systems engineering.

## Perspectives

Chapters 12 through 18 reflect an attempt to grapple with some of the implications of success, including the hazards this may present. Gregory Fahy considers some possible medical consequences of progress toward nanotechnology. Bill Joy presents a computer scientist's view of the utility of genuinely enormous computer power. Gordon Tullock presents an economic perspective. Ralph Merkle points to areas of risk that can be avoided. Lester Milbrath considers potential environmental impacts. And Arthur Kantrowitz discusses strategies for handling new technologies in an open society within a competitive world. Chapter 18 concludes this section with a panel discussion on the problems of public policy.

Two appendixes are also included for background. The first is an introductory essay on nanotechnology that I wrote for the *1990 Yearbook of Science and the Future,* published by Encyclopedia Britannica, "Machines of Inner Space." The second appendix is a presentation Richard Feynman made in 1959, "There's Plenty of Room at the Bottom: An Invitation to Enter a New Field of Physics," that demonstrates Feynman's characteristic insight. This talk was first published in the February 1960 issue of the California Institute of Technology's journal, *Engineering and Science.*

## Why Now?

Since the attempt to anticipate the consequences of nanotechnology may seem premature, some background is in order. Molecular systems engineering is a near-term prospect and the focus of this volume. Ultimately, however, molecular systems engineering will lead to an industrial infrastructure based on molecular manufacturing. Molecular nanotechnology, as used in this volume, does not refer merely to the fabrication of nanometer scale structures, but rather to a set of capabilities that will give thorough and inexpensive control of the structure of matter. If this is a real prospect, perhaps it is not too early to consider what happens when we succeed.

If there is some error in the arguments that point to this conclusion—that molecular manufacturing will ultimately give these powerful capabilities—I urge that these errors be vigorously and clearly identified. If there are no such errors, then I would argue that in the long run we all have a lot at stake. If molecular nanotechnology is possible, then large-scale applications are possible. Advanced nanotechnology will be based on highly productive systems of molecular machinery, which can include self-replicating systems. It is self-replicating systems of molecular machinery that are responsible for such prominent features of the earth as its forests and the composition of its atmosphere.

At this point, I would like to explain some of my own motivation for studying this field over the last dozen years and for giving talks on the conclusions. This seems to confuse some people in a way that hinders communication, so please indulge me for a moment. I regularly encounter people who assume that I must be seeking research grants for taking the next steps in the direction of nanotechnology. They then conclude that I am failing miserably because I have looked almost exclusively at developments that seem far beyond any reasonable research and development time horizon. The fact that I have never even submitted a grant proposal suggests, however, that this has not been the goal of my efforts. In fact, my major concern in past years has been to maximize the ratio of understanding of important long-term possibilities to actual progress toward those possibilities. My goal has been to stimulate a productive discussion of long-term prospects to give us more time for consideration of their consequences.

However, I have recently become persuaded of the following counterintuitive proposition: If one wants to maximize the time available for serious policy debate regarding a technology, it is best to start serious development efforts for that technology as soon as possible. Serious debate (it seems) will begin only with serious development efforts, and the earlier those begin, the worse the initial technology base will be. Therefore the time between the moment at which a serious debate begins and the moment at which an operable technology arrives will be longer with an early start than it would be if serious development were postponed until the technology base is more mature. In short, an early start means a longer debate and perhaps a wiser conclusion.

Thus, I am pleased that this volume includes contributions that consider the long-term policy implications of nanotechnology.

## Nanotechnology and the Cultures of Science and Engineering

The major challenge in developing the field of molecular systems engineering (and so getting an earlier start) is largely one of culture and organization. These challenges stem from the relationship between science and engineering. An experimental scientist (that is, one developing new techniques) is someone who solves engineering problems in order to advance human knowledge. An engineer (that is, one breaking new ground) is often someone who does science in order to advance capabilities. In chemistry, moving from a narrowly scientific stance to the goal of building complex molecular systems will bring a shift from synthesizing naturally occurring biological products and other small molecules to an effort that is arguably more creative: the combined design and synthesis of molecular components that can be part of large, wholly novel structures.

When engineers build large systems, they usually define their overall goals not in terms of the parts of the system but in terms of the capabilities of the system as a whole. They often seek the best understood and most easily fabricated parts, rather than the most puzzling phenomena. They focus on modular designs and usually work in coordinated teams. Large projects are broken into many pieces; many people work independently on the individual pieces, but they exercise their creativity with an understanding that all the pieces must fit together in the end. A complex problem is solved not by a single laboratory, but rather by as many teams as necessary to work on all the component problems—that is, to make all the needed components.

I recently asked two chemists at MIT for an example of a large collaboration in organic synthesis. After some consideration, they gave me one: The synthesis of vitamin B12 was divided between two labs, each of which synthesized half of the molecule. In engineering, I replied, an example of a large collaboration is the Apollo program. In the aerospace world—and in building particle accelerators—people have learned how to tackle large problems by dividing them into small pieces to be solved individually and then composed into a solution for the whole. I believe that we need at least medium-sized collaborations, as an engineer would see them, if we are to launch an effective effort in molecular systems engineering.

It has been suggested that I use the metaphor of the blind men and the elephant to illustrate pieces fitting together to form wholes. But this is a scientific metaphor: the elephant exists as part of nature, and an understanding of its parts will almost spontaneously form a coherent whole. Our situation, however, is more that of a community of people with diverse knowledge and skills (aerodynamics, metallurgy, making engines) who aim to bring these talents together to make an automobile, and then aim to learn to reduce the weight, add wings and a propeller, and build an aircraft. The need is not to put together pieces of knowledge to understand a coherent, existing thing, but rather to put together pieces of knowledge and ability to build working systems. This requires design and coordination.

At present, some of the most significant work, from a molecular systems engineering perspective, is being done in the engineering of proteins and other biomolecules. Curiously, these activities are often described and funded as "branches of biochemistry," that is, as a part of a study of nature. This may seem plausible today, but where would we be today if research in aeronautical engineering had been described and funded as a "branch of ornithology"?

As an example of the resulting problems, consider two descriptions of engineering. In the MIT School of Engineering course catalogue, engineering is defined as "a creative profession concerned with developing and applying scientific knowledge and technology to meet societal needs." On the other hand, in the "Instructions to Authors" of the journal *Protein Engineering*, engineering has quite a different meaning. "The objectives of those engaged in this area of research [protein engineering] are to investigate the principles by which particular structural features in proteins relate to the mechanisms through which the biological function is expressed, and to test these principles in an empirical fashion by introduction of specific changes followed by evaluation of any altered structural and/or functional properties." Period. In short, the purpose of protein engineering is to do science, to gain knowledge of the natural world. I believe that we will see rapid progress in the field of molecular systems engineering when there exist highly respected journals of protein engineering (and so forth) in which the instructions to authors unabashedly state that the purpose of molecular engineering is to make useful molecular devices, to expand human capabilities and to meet societal needs.

## Discipline in an Interdisciplinary Field

Nanotechnology is an interdisciplinary field. Sometimes work in such a field manages to escape from any discipline whatsoever. To avoid that fate for this field, I ask that you be harshly critical of anything that you hear called "nanotechnology"—my own words included.

# Notes

1. J. Foster, J. Frommer, P. Arnett, "Molecular manipulation using a tunnelling microscope," *Nature* 331 (January 1988): 324–326. See also J. Pethica, "Scanning tunnelling microscopes: atomic-scale engineering," page 301 of the same issue.
2. See R. Lerner, A. Tramontano, "Catalytic antibodies," *Scientific American* 258 (March 1988): 58–70.

# I

## Molecular Systems Engineering

# 2

# Atomic Imaging and Positioning

John Foster

Since the invention of the scanning tunneling microscope (STM) just a decade ago, the world has seen a revolution in the imaging and manipulation of individual atoms and molecules. This chapter describes some of the progress that has been made in visualizing and discriminating atoms on surfaces and in visualizing and performing chemical operations on individual molecules.

The group I work with at IBM is called Molecular Studies for Manufacturing. What I do and how I got into scanning tunneling microscopy deserves a word of explanation. We are primarily in the business of information storage, and storage during the last few decades has meant magnetic disk drives. These consist of a disk, some magnetic material to write on, and a head that flies over the disk to read and write data. Our current product uses a monolayer of small molecules (perfluorinated polyethers) as a lubricant on the thin film of the disk. As we push toward higher information densities, everything gets progressively smaller. The head is now less than 2000 Å (200 nm) from the disk, which moves at 60 to 80 miles per hour. To continue to improve these devices, we are trying to understand how molecules behave with respect to surfaces: how these molecules are oriented, how easily they move about, and so on.

## How the STM Works

This chapter describes work we have done with the scanning tunneling microscope (STM). The STM was invented in the early 1980s by Rohrer and Binning in IBM's Zurich laboratory, and they received the Nobel Prize for it.[1] What is stunning about the STM is its simplicity and its low cost. In general, it works by bringing a small conducting probe up to a conducting surface. In principal, no current flows between two conductors until they are touching. However, if the probe comes very close to the surface (less than 10 Å), very small amounts of current are produced. This happens because the electrons in the probe and the surface have wave functions (as described by quantum mechanics) that extend out into the vacuum. To the extent that these spillover wave functions overlap, a measurable current results. The amazing part about this current is that it depends exponentially on the spacing between the two conductors (as well as the voltage). This means that the current can increase by as much as a factor of 10 when the intervening space decreases from 5 to 4 Å. Thus the instrument can easily detect changes in spacing of a fraction of an angstrom.

To use the STM as an imaging tool, the probe is moved close to a surface, then the probe scans laterally across the surface while measuring the current. As the surface moves up and down, the current moves up and down as well. In practice, because the spacings are so extraordinarily small, a feedback loop keeps the current constant (e.g., at half a nanoampere) by moving the probe tip up and down. Thus the tip of the probe essentially traces over the atomic bumps on the surface, and this trace is transformed into an image.

Figure 2.1 shows how the probe is controlled by piezoelectric materials. These ceramic materials contract or expand when a voltage is applied to them. They tend to be very stiff so that voltages can be applied to move the tip of the probe with subangstrom accuracy. The technology to do this is nontrivial, but it has been known for years.

The instrument itself is a small device that can be held in the palm of a hand. It sits on several metal plates interspaced with rubber for vibration isolation. Because the tip and substrate are separated by only 1 to 10 Å, one might worry that ordinary movements in the room would crash the tip into the substrate. The rubber between the plates helps with vibration isolation, but the major factor in making a successful device is making the STM small and very stiff so that the tip and substrate move as a unit and not relative to each other. Also, the feedback mechanism reacts with a frequency of a few kilohertz while vibrations transmitted through the vibration-isolation mounts are typically only a

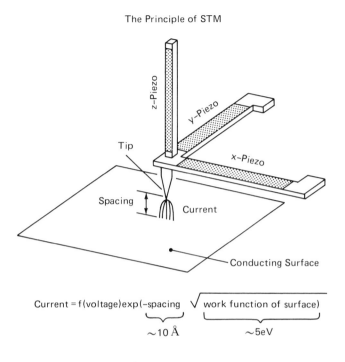

The Principle of STM

Current = f (voltage) exp (–spacing $\sqrt{\text{work function of surface}}$)

$\underbrace{\qquad}$ ~10 Å   $\underbrace{\qquad}$ ~5eV

Note: Spacing ↓ by 1 Å, current ↑ x 10

**Figure 2.1**
The principle of scanning tunneling microscopy (STM).

few hundred hertz. The entire device is often covered to isolate it from higher frequency vibrations transmitted through the air.

Because of the exponential dependence of current on separation, it turns out that the tip need be only cursorily sharpened—a radius of curvature of thousands of angstroms will do. This is easy to achieve by chance, and the first tips were made with a belt sander. It is only necessary for one atom to stick out an angstrom more than the rest for virtually all of the current to go through that one atom.

**Imaging Surface Structure with the STM**

The STM was first used to study surface science problems. Figure 2.2 shows the surface of silicon 1-1-1. The white balls are individual atoms

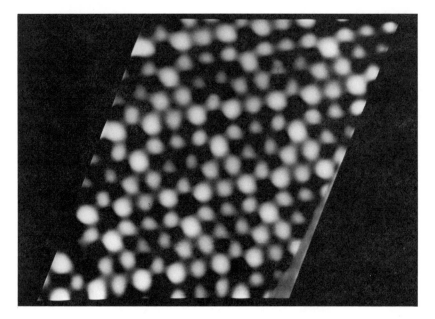

**Figure 2.2**
Silicon surface. (Courtesy Sung-il Park, Stanford University.)

of silicon in the 7 × 7 structure. Although this structure was anticipated by other techniques, the special feature of the STM is that it allows us to see individual atoms.

Plate 1 shows a surface of indium phosphide, one of the III-V materials discussed by Federico Capasso (chapter 8). Not only are individual atoms visible but they also have been computer processed to add false color to distinguish the indium and phosphorous atoms. Distinguishing different types of atoms is more complex than just tracing the bumps to get the geometry of the surface. By switching the direction of the current, the type of charge on each atom can be determined.

Another example is shown in the two panels of figure 2.3, which compare the same surface at the same time. The difference between the two panels of figure 2.3 is that the voltage between the tip and substrate has been reversed: in one case the electrons are entering the substrate and in the other they are leaving it. These pictures were taken at the same time by flipping the polarity and keeping track of the data. Two

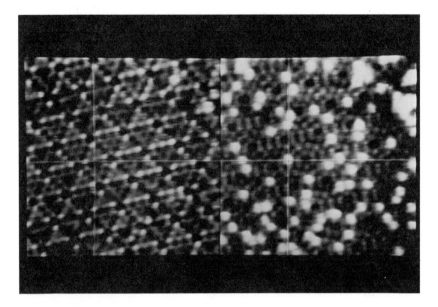

**Figure 2.3**
Two images of indium atoms on a silicon 7 × 7 surface created with opposite polarities of the voltage between surface and tip. (Courtesy Jun Nogami, Stanford University.)

pictures of indium phosphide were taken (using flipped polarity) and then combined to create plate 1.

Clearly, STMs do more than describe simple atomic topography. Figure 2.4 shows an energy diagram for how tunneling works. Energy is represented by the vertical dimension and space by the horizontal dimension. The Fermi level of the tip is shown at the left (there are electrons at this level and at lower energy levels). The sample is shown at the right. Imagine that the sample is a semiconductor, such as indium phosphide. If the bias is such that the tip is negative, the electrons leave the tip and move toward the sample. In the case shown in the middle portion of figure 2.4, in which the Fermi level of the sample is lower than the tip, electrons tunnel into empty energy states in the sample. In some sense, this measures the empty states of the sample. But if the polarity is reversed so that the tip is positive, electrons tunnel from the filled states of the sample into the tip, thus measuring the filled states. To the extent that the density states in the sample are known (i.e., the

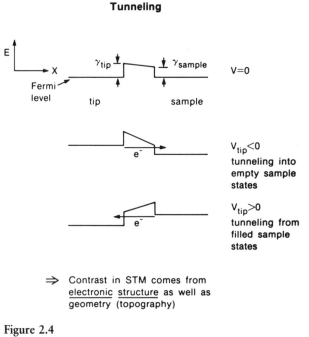

**Figure 2.4**
Energy level diagram for tunneling between surface and tip.

electronic structure), you can determine which kind of atoms you are looking at just by flipping the polarity back and forth. Because indium and phosphorous are in different families of the periodic table, they have different electron configurations and can be easily distinguished. The main point is that the images produced by the STM come from the electronic structure as well as the geometry of the sample.

### Imaging Molecules Adsorbed to a Surface with the STM

At IBM, Almaden, we wanted to explore new territory, so we decided to look at molecules adsorbed to a surface. We should have started by looking at molecules that behave like the solids in figure 2.3. The nice thing about imaging solids is that all of the atoms are more or less where you expect them to be. If you look at a surface of silicon atoms and they have the textbook arrangement and spacing, you can be pretty sure you're imaging silicon. But if you look at a molecule on a surface and

## Liquid Crystal Phases in the Bulk

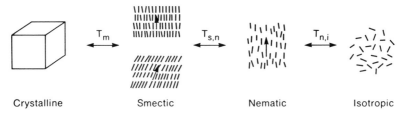

Figure 2.5
Liquid crystal phases.

see a blob, it's difficult convincing people that you're seeing a molecule and not just a piece of dirt. So eventually we decided to look at molecules that were ordered adsorbates on a surface, not unlike the ordered surfaces discussed above.

The class we chose was liquid crystals. These are long, thin molecules. At high temperatures they are isotropic, randomly oriented, as are typical liquids. As they cool, they tend to self-assemble and lie parallel to one another. This is called the *nematic* phase (figure 2.5). They have no order other than that they are all pointed in the same direction. This orientation is due to the fact that liquid crystal molecules consist of an aliphatic portion and an aromatic portion, and following the principle that similar groups like each other, the former all line up together as do the latter. This affinity constrains neighboring molecules geometrically so they all point in the same direction. If the liquid crystal is cooled further, some of the molecules enter the *smectic* phase, where they not only all point in the same direction, but they also line up in rows because the benzene rings tend to pack closely with each other, as do the aliphatic chains. In the smectic phase, different liquid crystals line up either in straight or tilted rows. If they are cooled still further, a crystalline solid is formed.

We put liquid crystals on a surface expecting that we would see the known spacing between molecules. There are many ways to prepare adsorbates of liquid crystals on a surface. We used a graphite surface because the freshly cleaved surface is atomically clean. Liquid crystal molecules are applied in a bulk drop onto the surface. The STM tip is brought into the drop, and the liquid wicks up around the tip so that

## Liquid Crystal Molecules

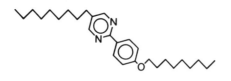

5-nonyl-2-n-nonoxylphenylpyrimidine (PYP 909)

Liquid crystalline order

**Figure 2.6**
Molecular structure of liquid crystal PYP 909.

the whole system is immersed in liquid crystal molecules. Because liquid crystal molecules are nonconducting, no current is measured until the probe gets very near to the graphite surface.

One of the first molecule we looked at was PYP 909 (figure 2.6). We expected this molecule to be ordered smectically with the molecules lying next to each other, with some characteristic spacing, and packed into rows. The STM picture (figure 2.7) shows an ordered array that does not look like a graphite surface. Our interpretation is that the large white bumps are the benzene rings and the smaller grey features are the aliphatic tails. We can even distinguish the angle that the benzene rings make with the aliphatic tails. The spacing between molecules is about 5.5 Å, just about as expected. This work, done with Jackie Spong, Jane Frommer, and others at IBM, Almaden, was the first unambiguous imaging of molecules in real space.

Visualization of these molecules occurs almost in real time. The output from the STM is fed into video at about 10 Hz, which is somewhat

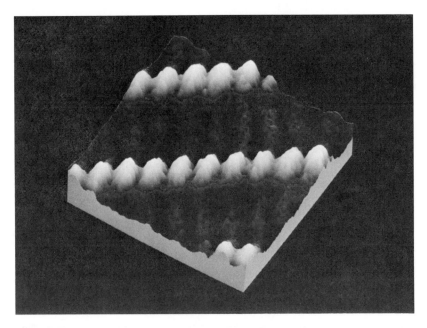

**Figure 2.7**
STM of liquid crystal PYP 909.

below normal video. We chose PYP 909 in particular for these experiments because we wanted to see phase transitions in real space and in real time. This molecule exhibits a transition from smectic C, where the molecules line up in rows of a tilted array, to smectic A, where they rotate to form rows of a vertically aligned array. We hoped to see an image like figure 2.7, raise the temperature, and watch the molecules rotate to form smectic A. Unfortunately, they did not undergo this phase transition on the surface. Instead, when we raised the temperature to 15° C above the isotropic temperature (so the bulk liquid was completely isotropic), the molecules adsorbed onto the graphite surface were still smectic C. They never rotated. At some point they just popped off and became isotropic.

We also used the STM to study the packing of several other organic molecules. For example, cyanobiphenyl molecules pack in pairs (figure 2.8a), but determining the exact structure of how they pack on a surface using other techniques is quite challenging. STM images permitted the

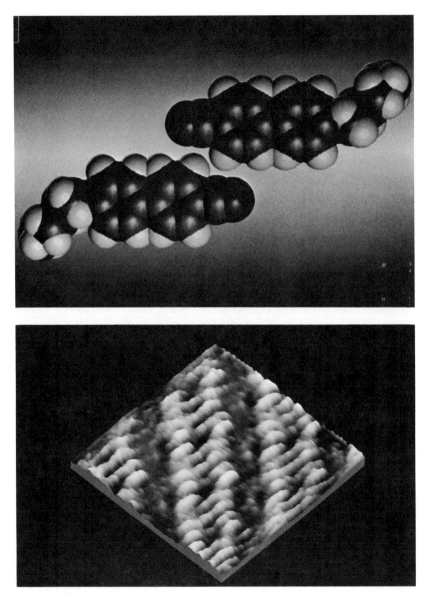

**Figure 2.8**
Molecular model of cyanobiphenyl structure (above) and STM of cyanobiphenyl
molecules on a surface (below).

visualization of the biphenyl portion and the aliphatic portion and the measurement of the angles and distances (figure 2.8b).

While creating these images, we wondered why the benzene rings look brighter than the aliphatic groups. The STM does not actually measure the topography of a molecule; it sees "topography" at the Fermi level of a conductor. The Fermi level of these molecules, or the highest filled molecular orbital, is volts below the level into which tunneling occurs. But this brightness means that there is more current. Did this mean that benzene rings are more conducting? Is this a feature of the aromatic electronic structure? We tried to answer this by substituting cyclohexane rings because cyclohexane has almost the same electronic structure as the aliphatic tails and is totally different from benzene. But bicyclohexane was still much brighter than the aliphatic tails. After many experiments using tunneling spectroscopy to figure out the electronic states, we convinced ourselves that the electronic states were not the source of the contrast in brightness.

What *is* the source of the contrast? The tunneling current depends on the barrier that the electrons have to go through. If the electrons go through a vacuum, that barrier is normally called the *work function,* which is the energy required to lift an electron out of the solid and into free space. The larger the barrier, the lower the current. In general, molecules can modify the barrier. Adsorbates on the surface can raise but usually lower the work function. The work function usually goes down because most molecules are polarized near the surface. The greater the polarizability of a molecule, the greater the decrease in the work function. The phenyl groups are more polarizable than the aliphatic groups, so the work function for the area under the phenyl groups is lower than under the aliphatic groups. In other words, the work function is locally modified by the different parts of a molecule. Benzene, biphenyl groups, and bicyclohexane groups have about the same polarizability—about twice the polarizability of aliphatic groups. We think this is the source of the contrast. In some sense, we don't really see molecules with an STM—we see the effect the molecules have on the substrate. This is also why we can get near-atomic but not quite atomic resolution when imaging molecules on a surface.

We still do not understand all of the physics involved in visualizing molecules on a surface. For example, would we be able to see a change in the work function brought about by a layer of 10 carbon atoms arranged vertically on a surface? Because the tip is 15 Å away (the height of 10 carbon atoms), the tunneling should be so small that no current would be seen. And a 10 carbon-long aliphatic tail is not conducting at these energy levels. The experiments are actually more complicated than this, and we and others have measured some current through conductance through those molecules. We do not understand why this phenomenon occurs. But in general, if the molecule is standing up on the surface, it does not lower the work function as it would lying horizontally on the surface. All the images presented in this chapter deal with a thin (3 to 4 Å) layer of molecules lying flat on the surface.

It is possible to measure molecules standing up on the surface, but it is a totally different phenomenon and is not well understood. For example, Langmuir-Blodgett films, which are standing up, have been studied, and they can be 50 Å thick for a lipid bilayer. This is much too thick to tunnel through, and it is supposedly not conducting, but researchers have imaged it with the STM. Some physical mechanisms that we do not understand are at work. Apparently the conductivity through these molecules is not simple.

### Manipulating Molecules on a Surface

After imaging a molecule, the next natural step is to do something to it. This work was done with Patrick Arnett and Jane Frommer.[2] We had voltage at our disposal, so that is what we used. A small voltage between the tip and the substrate is used to measure the current, so we decided to see what a large voltage would do. We placed a liquid, dimethyl phthalate or di(2-ethylhexyl) phthalate (which are not liquid crystal molecules) on to an atomically flat surface of graphite. In general, when you image the surface, none of these molecules appear—they are moving around rather than being adsorbed to the surface. Therefore, only the graphite atoms are imaged. We applied a narrow (100 ns) voltage pulse and discovered a threshold voltage: under 3.5 V, nothing happens. Over 3.5 V, we pinned a molecule to the surface (figure 2.9). The "bump" in figure 2.9 is about 8 Å across and the right general shape to be dimethyl phthalate, but we can't be sure because the resolution is not good enough to distinguish individual atoms.

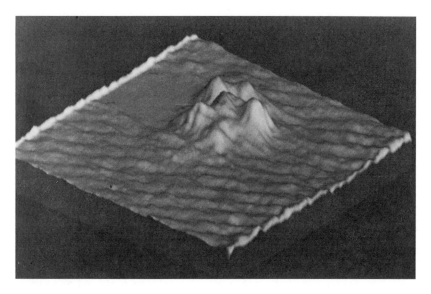

**Figure 2.9**
STM image of a dimethyl phthalate molecule pinned to the surface of graphite by a voltage pulse to the tunneling tip.

A molecule pinned to the surface can be removed from the surface with another voltage pulse. The probability of something happening as a function of the pulsed voltage is shown in figure 2.10 for both writing (pinning a molecule to the surface) and erasing (dislodging a pinned molecule to restore an atomically perfect surface). Erasing is easier than writing.

Looking at other pictures of pinned molecules reveals that they are not all identical. Thus the process is not reliable. In fact the bump may not be a single molecule. It may be a fraction of a single molecule, or fragments of two molecules that have come together on the surface. We think that some bumps represent a fraction of a molecule because a very interesting phenomenon is seen perhaps 15 percent of the time when a pulse is used to try to erase a molecule from the surface. In these cases, the bump is not completely released from the surface; part is released but part remains. In some cases we could whittle the molecule down two or three times and were left with a very small spot. In figure 2.11 the spots represent alternate carbon atoms on the graphite surface, 2.5 Å apart, and the larger triangular spot is a fragment of a molecule and

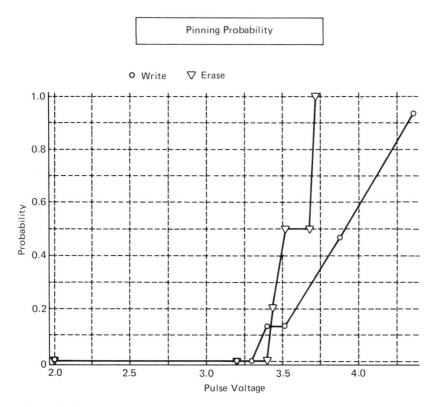

**Figure 2.10**
Probability of pinning or removing a molecule on the surface of graphite as a
function of pulse voltage applied to the tunneling tip.

is the smallest feature that anyone has ever made, on a localized scale,
with any tool.[3]

Another experiment was done by Mark McCord and collaborators at
IBM, Yorktown. It is an application of the STM to lithography that is
another twist to manipulating molecular systems. The STM tip is sur-
rounded by a gas, tungsten carbonyl ($W(CO)_6$), which is an unstable
molecule that tends to decompose when excited by a light beam or an
electron beam. In a manner analogous to schemes that use a laser to
write and deposit metal, he uses an STM, with a potential of 25 V at
the tip, drawing about 20 nA of current, to write little spots. The
tungsten is deposited on the surface and the CO escapes. (The spots turn
out to be tungsten carbide through reaction with the surface.) Figure

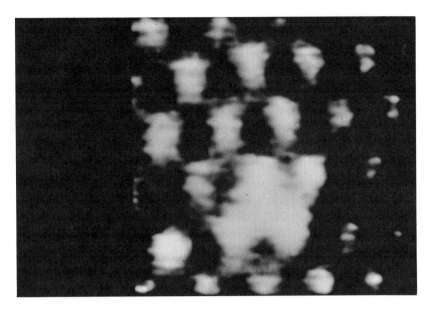

**Figure 2.11**
A small mark remaining on a graphite surface after the application of a voltage pulse to a pinned phthalate molecule.

2.12 is an SEM (scanning electron microscope) picture of dots written using an STM.

As the tip deposits the tungsten, it can still respond to the feedback loop that controls the tip position. As the tungsten deposit grows up toward the tip, the tip can pull back to keep a constant distance from the deposit. McCord has been able to grow a tall pillar, with an aspect ration of 10:1, as shown in panel B of figure 2.12. The one-dimensional "line width" of this column is about 300 Å. McCord has already extended this technique to make a line, shown in figure 2.13. He created what I would consider the first device made with an STM—a resistor. McCord can draw a line between two conductors that have been previously laid down and then measure the resistance across this STM-made wire.[4]

The letters shown in figure 2.14 were written at Stanford (by Moris Dovek, Tom Albrecht, Mike Kirk, and Chris Lang). These beautiful little letters are related to the work that I described earlier using voltage pulses on graphite, but instead of using 100 ns pulses, they use pulses of

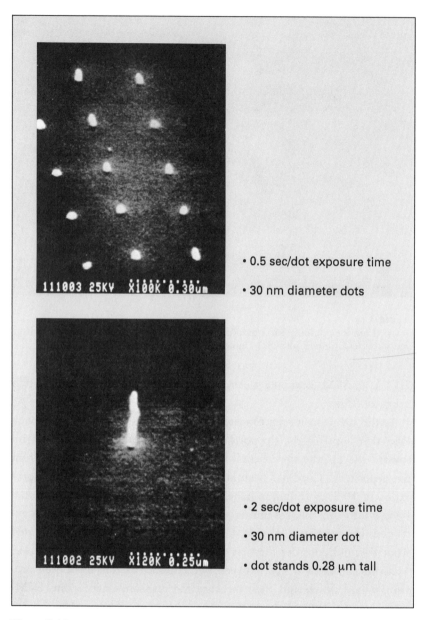

- 0.5 sec/dot exposure time

- 30 nm diameter dots

- 2 sec/dot exposure time

- 30 nm diameter dot

- dot stands 0.28 μm tall

**Figure 2.12**
Scanning electron microscope images of tungsten carbide dots (left) and a tungsten carbide column (right) produced by the controlled decomposition of $W(CO)_6$ using an STM.

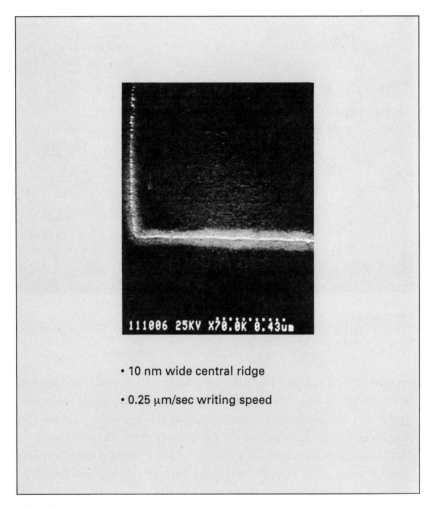

**Figure 2.13**
Scanning electron microscope image of a tungsten carbide line produced as in figure 2.12.

**Figure 2.14**
"STANFORD UNIVERSITY APRIL 1989" written using an STM to produce holes in a molecule-covered graphite surface.

microseconds duration. These longer pulses can actually rip graphite atoms out of the substrate. Our work (pinning and unpinning molecules) always left the substrate unperturbed, but the Stanford workers tore little holes one (3 Å) or two (6 Å) monolayers deep. This only works if the graphite surface is covered with molecules (it will not work in a very hard vacuum), so it is still a molecular interaction, but one driven by energetic electrons from the STM. The holes in figure 2.14 are about 20 Å in diameter. The technique is clearly reliable, since they were able to place the holes with sufficient precision to form letters. Thus you can use the STM not only to see atoms and molecules, and to manipulate molecules on the surface, but also to manipulate atoms on the surface in the presence of molecules.

**Future Directions**

We think that we can chemically manipulate molecules through interaction with energetic electrons. While the theoretical understanding pro-

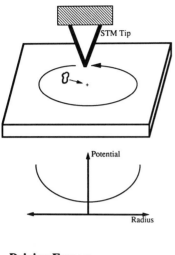

**Driving Forces:**

- Mechanical ———⊐→ *Spacing*
                 └→ *Trajectory*

- Electrical ———⊐→ *Voltage*
              └→ *Field*

**Figure 2.15**
Schematic of molecular herding.

gresses, we're working to be able to drive a molecule around on the surface from point A to point B. Some molecules that adhere only lightly to the substrate can be moved around with the STM in real time. As the tip moves closer to the system, molecules generally scurry away. This observation spawned the idea of "molecular herding," which I'm pursuing with Moris Dovek (figure 2.15).

I have a Shetland sheep dog, and they work by being generally annoying and circling the animal to be herded, causing it to move toward the middle. The idea is to surround the molecule with a special trajectory of the tip and presumably drive it in a certain direction. To move a molecule, you would move the whole trajectory, causing the molecule to move with it. This technique would create a potential for the molecule in a system as a function of the radius of the scan so that the potential is high at the outside and low in the middle; the molecule would generally sit in the middle.

The simplest approach is to physically tap the molecule and then adjust the spacing and the trajectory of the tip to achieve the desired potential shape. A more complicated technique would be to change the electrical potential of the tip as it moves and use that energy to move the molecule. As a first step toward molecular herding, we have used a spiral trajectory, instead of the usual raster trajectory, to do the scanning with the STM tip. We can use digital electronics to make any desired trajectory, potential, or spacing. Thus we should be able to drive any molecule toward the center and then drive it around at will.

**Notes**

1. G. Binning, H. Rohrer, "The scanning tunneling microscope," *Scientific American* 253 (August 1985): 50–56. The inventors of the STM describe its workings.

2. J. Foster, J. Frommer, P. Arnett, "Molecular manipulation using a tunneling microscope," *Nature* 331 (January 1988): 324–326. This paper is also the subject of an editorial by J. B. Pethica on page 301 of the same issue, "Scanning tunnelling microscopes: atomic-scale engineering."

3. D. Eigler, E. Schweizer, "Positioning single atoms with a scanning tunnelling microscope," *Nature* 344 (1990): 524–526. Less than six months after this conference, another group at IBM, Almaden, reported positioning single xenon atoms on a nickel crystal surface at low temperatures (4 K) and used the STM to write the letters "IBM" using 35 xenon atoms.

4. Recent reports of using the STM to demonstrate switching properties for electronic devices and to perform nanolithography include:

I. Lyo, P. Avouris, "Negative differential resistance on the atomic scale: implications for atomic scale devices," *Science* 245 (1989): 1369–1371.

P. Bedrossian, D. Chen, K. Mortensen, J. Golovchenko, "Demonstration of the tunnel-diode effect on an atomic scale," *Nature* 342 (1989): 258–260.

E. Garfunkel, G. Rudd, D. Novak, S. Wang, G. Ebert, M. Greenblatt, T. Gustafsson, S. Garofalini, "Scanning Tunneling Microscopy and nanolithography on a conducting oxide, $Rb_{0.3}MoO_3$," *Science* 246 (1989): 99–100.

**Discussion**

*Audience:*   How fast do you move the probe?

*Foster:*   Typically we scan at many kilohertz so that it takes about 0.2 milliseconds to make a loop.

*Audience:*   Any thoughts on the mechanism of pinning molecules to the graphite surface?

*Foster:*   The graphite surface is totally inert; you don't see any reaction at all with the surface unless you perturb it. The thought that we (and others) have had is that a surface atom in the totally planar graphite surface pops up very slightly. Every atom of graphite is trigonally bound (all of its bonds are in the plane), but if you force it up slightly, it becomes tetrahedrally bound so that it has a very reactive bond pointing straight up. Thus if you dump energetic electrons into an antibonding orbital, it pops the atom up ever so slightly. Then the graphite carbon atom grabs a molecule, if one is there to be grabbed, and bonds to it. If no molecule is there, the carbon atom simply drops back down into the surface and nothing happens. If you dump a large number of electrons over a large surface, you can pop up the whole surface. Then you can make the whole surface pop off. This is like turning graphite into diamond.

*Audience:*   What about thermal effects?

*Foster:*   From mundane general physics arguments, the temperature rise expected from the power of the pulses that we are using is really small, only a fraction of a degree centigrade. But this makes assumptions about the mean free path of the electron, where the energy is dissipated, and so on, that could be wrong. Also, in our experiments we see a threshold in the energy (voltage) applied, but not in the amount of power dumped into the system. So we don't believe that thermal effects are important here.

*Audience:*   Could you use an STM to make disk drives and, if so, when?

*Foster:*   The information storage industry has to use very reliable systems. Using STMs for information storage is very far in the future. In the meantime we are using the STM to understand molecules on surfaces, which is very relevant to our business. However, there are several groups that are thinking about storing information with STMs.

*Audience:*   There has been a project for several years now at Lawrence Livermore Labs using a laser to create circuit elements by pyrolytic decomposition. You can use water and oxygen to deposit oxide on silicon and do everything necessary to build integrated circuits. Is anyone going in that direction with the STM?

*Foster:*   You're talking about really interesting work by Bruce Mc-

Williams. The work by McCord that I mentioned using tungsten carbonyl falls along similar lines. The answer to your question is yes. In
my paper I didn't discuss the electrochemistry at all. You can try to
modify surfaces in electrochemical solution. Several groups are working
on this; there are many exciting possibilities.

# 3

# Design and Characterization of 4-Helix Bundle Proteins

Tracy Handel

## De Novo Design of α-Helical Proteins

Despite the large library of proteins for which both sequence and structural information exists (via crystallography or nuclear magnetic resonance), it is still not possible to predict the three-dimensional structure of a protein based on sequence information alone, at least not in the absence of sequence homology comparisons to proteins of known structure. Current strategies to define the rules governing protein folding include both computation and experimental approaches (i.e., mutagenesis). An alternative approach is de novo protein design where one seeks to design a primary sequence that will fold into a protein of predefined tertiary structure. This chapter describes how we in the DeGrado group at du Pont have been trying to design proteins that have folding motifs similar to those found in naturally occurring proteins, but which have much simpler peptide sequences.[1, 2]

## Building Proteins from Degenerate Units of Secondary Structure

In a protein-design-based approach to protein folding, we still face the difficulty of relating a primary sequence of amino acid residues to a tertiary (three-dimensional) structure. What is more tractable, however, is predicting sequences that give rise to units of particular secondary structure such as α-helices or β-sheets. We can therefore try to use this knowledge base to design repetitive units of secondary structures that have packing complementarity and can therefore self-associate in such a way as to give rise to a predefined tertiary structure. The α-helix is a good candidate for this approach because we know many rules governing helix formation, and we can make several different kinds of proteins based solely on α-helices. In our group there are three classes of proteins, all involving α-helices, in which significant progress has been made:

• α-helical coiled coils.[3] The peptide backbone of an α-helical coiled coil is shown in plate 2.[4] This is an example of a folding motif found in some gene-regulatory proteins that bind to DNA such as Fos and Jun.[5]

• Ion channels. A view down the pore of a cation-selective ion channel that has been designed by our group is shown in plate 3.[6] These structures consist of aggregates of several simple α-helices that insert into membranes and conduct ions upon application of a membrane potential. The ion selectively of these channels depends in part on the size of the opening and therefore on the aggregation state of the peptides.

• Four-helix bundle proteins. The 4-helix bundle protein shown in plate 4 is a folding motif found in many proteins, including myohemerythrin and cytochrome c'.

Four-helix bundle proteins are the focus of this chapter; I will describe the principles that were used to design the four-helix bundle,[7] approaches using nuclear magnetic resonance (NMR) to determine how close the actual structure is to that anticipated by the design and predicted by modeling, and recent attempts to increase the functionality of the bundle with metal-binding sites.

The immediate goals of this work are threefold:

1. Understanding helix formation. A great deal is already known about helix formation, particularly due to contributions from several labs (including the work of Scheraga, Kallenbach, Baldwin, and DeGrado).[8] However, there remains much to be learned about the relationship between the primary sequence of amino acid residues in a polypeptide chain and the ability of such sequences to fold into a-helices.

2. Understanding helix-helix packing. In the course of this work we have found that predicting the way in which helices pack against one another to give rise to a particular tertiary structure is perhaps a more challenging problem than predicting whether or not a given peptide sequence will adopt a helical conformation. However it is a crucial part of de novo design and protein folding.

3. Designing functional helical bundles. Eventually we want to design proteins that are not simply framework structures but exhibit particular

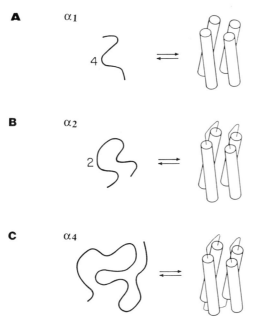

**Figure 3.1**
Incremental strategy for designing the 4-helix bundle.

functions. For example, in the case of the 4-helix bundles, we are interested in making metal-binding 4-helix bundles that can perform interesting catalytic or redox chemistry, or bundles that can fold and unfold in response to chemical stimuli.

Ultimately we want to understand more than just how to make helical proteins. A Holy Grail of structural biology is to understand all of the determinants of protein folding so that we can create proteins that are complex, multidomain, and impressive in functionality.

## Strategies in the Design of a 4-Helix Bundle Protein

The incremental approach that was taken to design the 4-helix bundle is shown schematically in figure 3.1 (see references 1 and 2). First a 16-residue peptide ($\alpha_1$B) was designed such that four peptides would self-associate cooperatively into an unlinked 4-helix bundle. In a second step, loops were inserted between the helices such that two helix-loop-helix pairs ($\alpha_2$) would dimerize to give a 4-helix bundle. In the last step, a third loop was added to generate a monomolecular 4-helix bundle ($\alpha_4$).

**Designing a 16-Residue Sequence of $\alpha_1$B**

There were several design principles used in generating the amino acid sequence for the 16-residue unlinked helices ($\alpha_1$B). First, it was assumed that hydrophobic interactions would drive tetramerization. Therefore, the helices were designed to be amphipathic, that is, one side being hydrophobic and the other side polar. Thus, when the peptide tetramerizes, the hydrophobic residues will be clustered in the bundle interior whereas the polar residues will be aqueous-exposed. Second, the conformational preferences of the individual amino acids to be in an $\alpha$-helix were considered. For example, lysine and leucine are amino acids with a high propensity to form $\alpha$-helices (according to various scales such as those by Scheraga and Chou Fasman[9]) so these were included in the sequence as hydrophilic and hydrophobic components, respectively. On the other hand, glycine tends to break helices so glycines were included at the two ends of the helix as helix-stops, in preparation for adding the loops. Third, electrostatic interactions are also important, both between the side chains of the amino acids and also the helix macrodipole.[10] Thus glutamic acids were included at the amino terminus and lysines at the carboxyl terminus to compensate for the helix dipole. Glutamic acid and lysine residues were also paired at positions i, i + 3 and i, i + 4 in order to form salt bridges for electrostatic stabilization.

Figure 3.2 shows the sequence of the 16-residue helix, $\alpha_1$B. In contrast to most naturally occurring proteins, the helix is extremely simple in composition and consists of only four amino acids: six leucines, four glutamic acids, four lysines, and two glycines. The "helical wheel" diagram in figure 3.2 shows the spatial distribution of these residues and the segregation of hydrophobic residues to one side of the helix and what would be the interior of the bundle and the hydrophilic residues to the other side, which will be surface-exposed in the bundle. Also, the glutamic acid and lysine residues are positioned along the helix in such a way that they have the potential to form salt bridges.

In concert with the design, computer modeling was used to predict the arrangement of the helices in a 4-helix bundle. First an array of four antiparallel $\alpha$-helices was created by application of a 2,2,2-symmetry

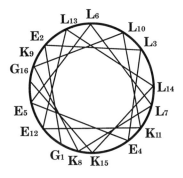

Ac-G-E-L-E-E-L-L-K-K-L-K-E-L-L-K-G-CONH$_2$

**Figure 3.2**
Helical wheel diagram and amino acid sequence of $\alpha_1$B.

operator to a single helix.[11] Subsequently the tilt and translation of the helices parallel and perpendicular to the bundle axis were manipulated to orient the leucine side chains inward and with effective interdigitization. The structure was then subject to iterative dynamics calculations and energy minimization using the AMBER force field. The resulting structure is that shown in plate 4. Features of the bundle that are suggested by the model include the antiparallel orientation of the helices and the offset of the helices for close packing of the leucine side chains in the bundle interior.

Once the design was complete, the peptide, designated $\alpha_1$B, was synthesized by solid-phase methods and characterized initially by size exclusion chromatography and sedimentation equilibria for molecular weight determination and by circular dichroism for helix content and aggregation state.[12] The molecular weight measurements are consistent with the formation of a tetramer at concentrations above 10 mm. The circular dichroism spectra show double minima at 222 nm and 208 nm, confirming the helical character of the peptide. Furthermore, the concentration dependence of the mean residue ellipticity at 222 nm ($\theta_{222}$) is best modeled as a cooperative monomer to tetramer equilibrium where the peptide is approximately 30% helical in the monomeric state and 75% helical in the tetrameric state (figure 3.3). Taken together these data support the conclusion that these peptides form 4-helix bundles.

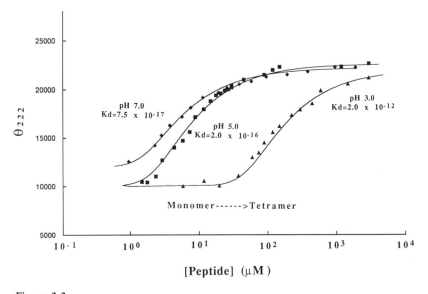

**Figure 3.3**
The dependence of $\theta_{222}$ on peptide concentration at pH 3.0, 5.0, and 7.0. The data were fit by theoretical curves describing monomer-tetramer equilibria as described in the text (see references 2 and 13).

Analysis of the data in figure 3.3 also allows calculation of the stability of the tetrameric complex. As might be expected from the sequence, the stability is pH-dependent and is greater at pH 5.0 and 7.0, where the glutamic acid residues are charged, than at pH 3.0, where they are protonated. This suggests the importance of the Glu-Lys electrostatic interactions on helix stability.

### Introducing Loops to Make a Dimeric 4-Helix Bundle Protein ($\alpha_2$) and a Monomolecular 4-Helix Bundle ($\alpha_4$)

The next step in the design was to insert a loop between two helices such that two helix-loop-helix pairs would dimerize to give a 4-helix bundle (figure 3.1, step B). Introduction of a proline between the C-terminus of one 16-mer and the N-terminus of another (helix-P-helix) resulted in the formation of a trimeric aggregate. This is presumably because the short length of the "loop" did not allow the two helical regions of the peptide to fold back on each other. However, a simple

**Table 3.1**  Modular design of a 4-helix bundle protein

| Helix | GELEELLKKLKELLKG |
|---|---|
| Loop | PRR |
| $\alpha_1 B$ | Ac-Helix-CONH$_2$ |
| $\alpha_2$ | Ac-Helix-Loop-Helix-CONH$_2$ |
| $\alpha_4$ | Met-Helix-Loop-Helix-Loop-Helix-Loop-Helix |

trimer (Pro-Arg-Arg) inserted between two helices was sufficient to initiate a turn and promote dimer formation.

In the final step of the design, a third PRR loop was introduced to generate a monomolecular 4-helix bundle protein. The sequences of the three constructs are summarized in table 3.1. The 16-residue peptide ($\alpha_1 B$) and the helix-loop-helix peptide ($\alpha_2$) should each self-associate to form a 4-helix bundle. The addition of a third loop produces the full protein ($\alpha_4$), which should by itself fold into a 4-helix bundle.

Both $\alpha_1 B$ and $\alpha_2$ were produced by solid-phase synthesis. In the case of $\alpha_4$, a synthetic gene was constructed and the protein was expressed in E. coli at fairly high levels. Table 3.2 summarizes the free energies of folding and aggregation states of these proteins. In all cases a 4-helix bundle was obtained as determined experimentally by gel filtration studies; $\alpha_1$ tetramerizes, $\alpha_2$ dimerizes, and $\alpha_4$ is monomeric. The free energies of folding shown in table 3.2 are calculated from these denaturation curves based on appropriate equilibrium expressions for the unfolding

**Table 3.2**  Summary of aggregation states and free energy of folding

| Peptide | $\Delta G$ (kcal mol$^{-1}$) | Aggregation number |
|---|---|---|
| $\alpha_1$ | $-18.6$[a] | 4.4[a] |
| $\alpha_2$ | $-12.8$[a] | 2.2[a] |
| $\alpha_4$ | $-22.5$[b] | .98 |
| Lysozyme | $-8.9$[b] | |
| Myoglobin | $-7.6$[b] | |
| Ribonuclease A | $-7.5$[b] | |
| $\alpha$-Lactalbumin | $-4.2$[b] | |

[a]See Ho and DeGrado, 1987.
[b]See Regan and DeGrado, 1988.

(monomer-tetramer for $\alpha_1B$ and monomer-dimer for $\alpha_2$). A comparison of the free energy obtained from folding these proteins to that of several small natural proteins indicates that these designed proteins are unusually stable. This may be due in part to the fact that they have a highly hydrophobic core, consisting exclusively of leucine residues. This inherent stability is important as it provides the opportunity for altering the protein sequence (e.g., by introducing metal-binding sites) without intolerable destabilization of the protein.

## Characterization of a 4-Helix Bundle by NMR

To summarize, based on circular dichroism (CD) data and gel filtration measurements, we know that $\alpha_1B$, $\alpha_2$, and $\alpha_4$ have the correct aggregation state, a high helical content, and an unusually high thermodynamic stability. What we do not know, however, is the high-resolution features of how the four helices are packed together to give a 4-helix bundle. If we want to extend our approach from engineering a framework structure to making a bundle with functionality, we need detailed structural information. Furthermore, if we are to use this approach to design proteins with a predetermined fold, we need feedback so that we can iteratively improve our designs.

To obtain structural information we have used one-, two-, and three-dimensional NMR.[13,14] Most of our initial work has been done on the unlinked bundle $\alpha_1B$ because it was expected to be highly symmetrical, which would facilitate resonance assignments in the NMR spectra. Furthermore, with $\alpha_1B$ we could incorporate isotopically-labeled amino acids at specific positions by solid-phase synthesis. This is not possible with $\alpha_4$ as it is expressed biosynthetically in E. coli.

### One-Dimensional Proton Exchange

Initially we used one-dimensional $^1H$ NMR to measure the rate of hydrogen-deuteron exchange for the amide protons of $\alpha_1B$. These experiments tell us which amide protons are involved in hydrogen-bonding because the exchange rates of hydrogen-bonded amides will be slow relative to nonhydrogen-bonded amides. Thus we can map out regions of the peptide chain that have a well-defined secondary structure. From these exchange experiments we calculated "protection factors," which

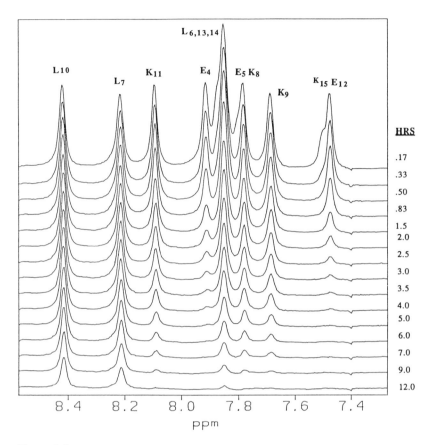

**Figure 3.4**
Proton-deuteron exchange of $\alpha_1 B$ at pH 5.0.

are a measure of how highly protected the amide protons are in the folded state relative to a theoretically calculated value for the unfolded state. Figure 3.4 shows the amide region of the proton spectra of $\alpha_1 B$ at various times after the peptide has been solubilized in $D_2O$.[15] From the intensity of resonances as a function of time, we calculated exchange rates for the individual amides. Qualitatively, several of the residues at the beginning and end of the helix are very rapidly exchanged. These include the first three residues, which is to be expected because the three N-terminal residues within any helix are not H-bonded. $Gly_{16}$ is also exchanged almost immediately, which suggests it may not be part of the helix. The remaining internal residues exchange relatively slowly, which

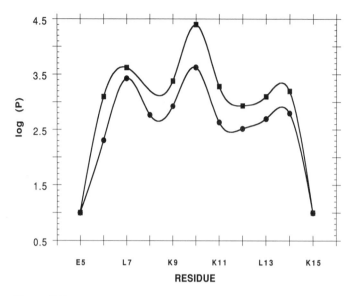

**Figure 3.5**
Log of protection factors for $\alpha_1B$ at pH 5.0 in the presence (■) and absence
(●) of 0.1 m NaCl.

implies that those amide protons are stabilized by H-bonding within the
helix.

One can convert the exchange rates to protection factors according
to the method of Molday.[16] The protection factor of a given residue is
the measured exchange rate as corrected for sequence, temperature, and
pH effects. Consequently protection factors more accurately reflect structural details (stability of the H-bonds and solvent accessibility) than the
uncorrected exchange rates. Values of the protection factors are plotted
as a function of residue in figure 3.5, and range from about 10 to 4200
in the absence of NaCl, and from about 10 to 25,000 in the presence
of 0.1 m NaCl.[17] Overall, the plot has a domed shape, and the highest
protection is observed for $Leu_{10}$ and $Leu_7$. It is interesting to note that
these amide protons are the most buried in the models of $\alpha_1B$ tetramer.
The termini of the helices show lower values of P, reflecting higher
solvent accessibility and flexibility. Superimposed on the dome is a periodic distribution of P, reflecting the $\alpha$-helical nature of the peptide and
the reduced exposure of leucine residues.

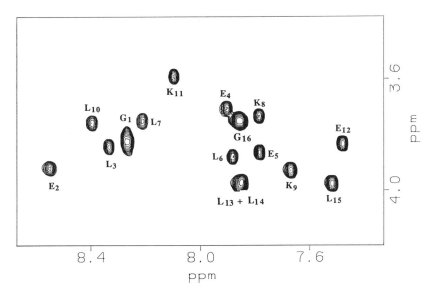

**Figure 3.6**
HOHAHA of the N$\underline{H}$-C$_\alpha$$\underline{H}$ region of $\alpha_1$B at pH 5.0, 20° C.

## Two-Dimensional NMR for Structure Determination

To probe more directly the structure of $\alpha_1$B, we carried out a series of standard two-dimensional experiments: double quantum-filtered correlated spectroscopy (DQF-COSY), nuclear Overhauser-effect spectroscopy (NOESY), and homonuclear Hartmann-Hahn spectroscopy (HOHAHA). From the combination of these three experiments we have been able to assign all proton resonances in the bundle.

Figure 3.6 shows the N$\underline{H}$-C$_\alpha$$\underline{H}$ region of a two-dimensional HOHAHA experiment. This is an instructive region of the two-dimensional spectrum, referred to as the "fingerprint" region, since every residue in the protein should contribute a cross-peak due to scalar coupling between the intraresidue amide N$\underline{H}$ and C$_\alpha$$\underline{H}$ protons.[18]

The different resonance positions of the cross-peaks reflect different chemical environments of the amide and C$_\alpha$ protons. What is interesting is that we observed only one cross-peak per residue in $\alpha_1$B, that is, 16 cross-peaks for the 16-residue peptide instead of 64 cross-peaks for the 64-residue bundle. Since the NMR spectra are recorded under conditions in which we have nearly 100% tetramer, these data indicate that every helix is spectroscopically identical. This may be due to the fact that the bundle has the 2,2,2 symmetry, as suggested by the model, or that motional averaging causes the helices to sample the same average magnetic environment on the NMR time scale. We do not know the answer to this question, but we are trying to address it.

The bread and butter experiment for NMR structure calculations is the NOESY experiment. In contrast to correlation experiments (i.e., COSY or HOHAHA), where cross-peaks arise from *through-bond* magnetization transfer, cross-peaks in a two-dimensional NOESY experiment arise from *through-space* dipolar interactions between protons that are separated by 5 Å or less. The NOESY experiment therefore gives us distance information between proton pairs and thereby allows us to define the secondary and tertiary structure of a protein.

To determine elements of secondary structure, we look for patterns of short- and medium-range NOEs. For example, in $\alpha$-helices, sequential amide protons are separated by ~2.5 Å, and therefore we expect to see N<u>H</u>-N<u>H</u> (i, i + 1) cross-peaks for the helical regions of proteins. In addition, we also expect cross-peaks between the amide proton of one residue and the $C_\alpha$ proton of the amino acid three residues distant in the sequence (approximately one turn of the helix). Table 3.3 shows a summary of the short- and medium-range NOEs. Based on the sequential N<u>H</u> and $C_\alpha$<u>H</u>-N<u>H</u> (i, i + 3) connectivities, it appears $\alpha_1$B is helical between residues 2–15. (For details, see reference 13.)

Once elements of secondary structure have been identified via patterns of short- and medium-range NOEs, one then attempts to identify long-range NOEs that define the way they are packed together to give the tertiary fold of the protein. In the case of $\alpha_1$B, we want to know how the helices associate: Do they have an antiparallel orientation? What are the crossing angles? And are the leucine side chains interdigitated or abutted? Despite the simplicity of the peptide, this final step of the structure determination has proven to be extremely difficult by conventional approaches because of the spectroscopic equivalence of the helices and the extreme overlap of the side chain resonances. If every helix is identical, there is no addressable information as to how the four helices interact. For example, NOEs between $Leu_6$ on helix 1 and $Leu_{10}$ on helix 2 are indistinguishable from *intrahelical* NOEs between $Leu_6$ and $Leu_{10}$ on any one of the four helices in the tetramer.

**Table 3.3**   Summary of short- and medium-range NOEs of $\alpha_1$B at 200 msec mixing time, pH 5.0.

| | | 1 | 2 | 3 | 4 | 5 | 6 | 7 | 8 | 9 | 10 | 11 | 12 | 13 | 14 | 15 | 16 |
|---|---|---|---|---|---|---|---|---|---|---|---|---|---|---|---|---|---|
| | | G | E | L | E | E | L | L | K | K | L | K | E | L | L | K | G |
| $J_{\alpha N}$ | | | 5.8 | 3.9 | 3.9 | <5.0 | <5.0 | 3.9 | <4.0 | <5.0 | 5.1 | | <4.0 | 4.3 | 5.6 | 6.1 | 4.5 | 6.4 |

As an alternative approach to obtaining some information about helix-helix packing, we prepared "mutants" of the $\alpha_1$B peptide. On the basis of the computer model of the 4-helix bundle, we selected leucine pairs that could be replaced by unique hydrophobic residues such that NOEs between them could be unequivocally attributed to interhelical NOEs. In the first variant, Leu$_3$ and Leu$_{13}$ were replaced with a valine and a phenylalanine, respectively. Based on our computer model, these residues should be close enough in space to give rise to NOEs if the helices align in an antiparallel orientation (plate 5). Our choice of valine and phenylalanine was prompted by the fact that the sum of the volumes of these two residues is approximately equal to two leucines, which would minimize packing distortions. Furthermore, the side chain protons of these two residues resonate in very different parts of the NMR spectrum, which helps alleviate the overlap problem. We also thought ring current shifts from the phenylalanine would increase the dispersion in the spectrum of this peptide relative to $\alpha_1$B.

The mutant peptide (Val$_3$Phe$_{13}$) has many characteristics similar to $\alpha_1$B. It associates cooperatively to form a tetramer and is destabilized in the tetrameric state by only a few kcal/mole relative to $\alpha_1$B. Also, the four helices within the bundle are spectroscopically identical. However, inclusion of the valine and phenylalanine residues does permit the identification of long-range NOEs between helices. Figure 3.7 shows a portion of a two-dimensional NOESY experiment of Val$_3$Phe$_{13}$.

This peptide was synthesized with uniformly deuterated leucine (so the leucines are NMR-silent) to eliminate ambiguities due to the fact that the valine and leucine methyl protons overlap. Thus the arrows highlight cross-peaks that arise unequivocally from dipolar interactions between the phenylalanine ring protons that resonate along the x axis and the valine $\beta$ and $\gamma$ protons that resonate at the ppm positions defined by the y axis. These data therefore indicate that there is a population of the helices that are oriented in an antiparallel fashion, consistent with the design. It should be noted, however, that we cannot exclude the presence of helices that are parallel because NOEs diagnostic of this orientation are not visible for symmetry reasons.

### Three-Dimensional NOESY-TOCSY Experiments on Isotopically Labeled Peptide

Although we have been able to obtain interhelical NOEs using Val$_3$Phe$_{13}$, the number is insufficient to use as constraints for distance geometry calculations of a three-dimensional structure. However, we are pursuing other approaches to obtain interhelical NOEs by isotopic labeling of the leucine residues. Val$_3$Phe$_{13}$ has four residual leucines (Leu$_6$, Leu$_7$, Leu$_{10}$, and Leu$_{14}$). Our intention was to synthesize four different peptides, each with only a single protonated leucine, all others deuterated. One can then prepare all possible 50:50 mixtures of the peptides. In a NOESY experiment on any one of these peptide mixtures, NOEs between the two protonated leucines can only arise from *interhelical* cross relaxation.

We have recently initiated these experiments. Unfortunately, two-dimensional NOESY spectra of the mixtures still suffer from insufficient spectral dispersion and identification of leucine-leucine NOEs is not possible. For this reason we have implemented three-dimensional NOESY-TOCSY experiments as described by Kaptein et al. and Oshkinat et al.[19] The concept of these experiments is to use the additional

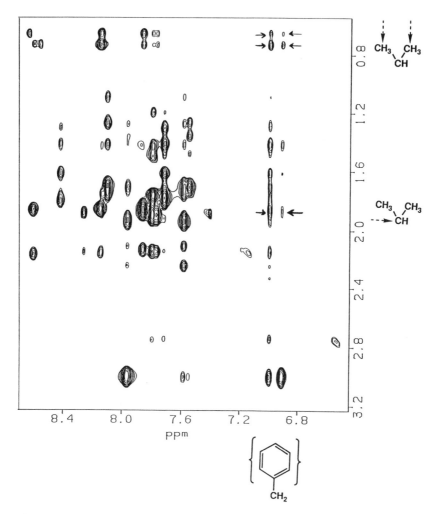

**Figure 3.7**
NOESY of Val$_3$Phe$_{13}$ at 150 msec mixing time, pH 5.0.

resolution gained by dispersing frequencies into a three-dimensional cube, coupled with selective deuteration of all but one leucine per helix. The NOESY portion of the experiment transfers magnetization to all spatially close protons by virtue of its distance dependence, then magnetization is transferred within a given resonance by the TOCSY (HOHAHA) step via J-coupling. In this manner structurally important NOE cross-peaks can be relayed to more well-resolved regions of the spectrum. The concept is illustrated with a 50:50 mixture of $Val_3Phe_{13}$ (H-$Leu_6$) + $Val_3Phe_{13}$ (H-$Leu_{10}$). The complete data set is contained in a three-dimensional cube. More informative, however, are two-dimensional slices of the cube. Plate 6 shows an overlay of the mixture with pure $Val_3Phe_{13}$ (H-$Leu_6$), and plate 7 shows an overlay of the mixture with $Val_3Phe_{13}$ (H-$Leu_{10}$). The cross-peak delineated by the arrow arises at the $F_1$, $F_2$ frequencies of the leucine methyls. Since this cross-peak *only* occurs in the mixture, it must arise from an NOE between the methyl groups of $Leu_6$ and $Leu_{10}$—which was expected to be present based on our model of this bundle.

Similar experiments will be carried out for all pairwise combinations of selectively deuterated peptides to obtain as much structural information as possible. It must be stressed, however, that if the spectral equivalence of the helices turns out to be due to motional averaging, we may not be able to get a complete three-dimensional structure. If we find that the structure is truly symmetric, we are well on the way to a complete structure. On the other hand, if the bundle is dynamic, we suspect the structures may be more like "molten globules," or intermediates in protein folding, rather than well-defined native states of proteins.[20] If this is the case, in order to engineer proteins with more nativelike characteristics, we may need to introduce interactions between the helices that are more specific than hydrophobic interactions. The introduction of metal sites, which have a strong dependence on distance and angular geometry, is our first step in this direction. It is also a critical step in designing functional helical bundles.

### Design of Metal-Binding 4-Helix Bundles

With the dual goal of engineering functional 4-helix bundles, as well as imparting more specificity in the helix-helix packing, we have introduced

Helix 1 :   G-E-L-E-E-L-L-K-K-L-K-E-L-L-K-G
Helix 2 :   G-E-L-E-E-L-H-K-K-L-H-E-L-L-K-G
Helix 3 :   G-E-L-E-E-L-H-K-K-L-L-E-L-L-K-G
Loop :      P-R-R

H2$\alpha_2$ :    Ac-Helix 1 - Loop - Helix 2 - CONH$_2$
H3$\alpha_2$ :    Ac-Helix 3 - Loop - Helix 2 - CONH$_2$
H3$\alpha_4$ :    Ac-Helix 1 - Loop - Helix 3 - Loop -
          Helix 2 - Loop - Helix 1- CONH$_2$
H6$\alpha_4$ :    Ac-Helix 3 - Loop - Helix 2 - Loop -
          Helix 3 - Loop - Helix 2- CONH$_2$

**Figure 3.8**
Sequences for the four metal-binding peptides.

metal-binding sites into the framework 4-helix bundles.[21] Starting with a model of $\alpha_2$, we substituted two histidines on one helix at positions i, i + 4 and a third histidine on the adjacent helix. The histidine residues could be positioned at three corners of a tetrahedron or octahedron by introducing energetically reasonable torsional angles into the side chain with solvent molecules serving as the remaining ligands. In our metal binding 4-helix bundle, designated H3$\alpha_2$, there are two metal sites at symmetry related positions in the bundle because $\alpha_2$ dimerizes to produce a 4-helix bundle. A similar single-site bundle (H3$\alpha_4$) and a two-site bundle (H6$\alpha_4$) were engineered from $\alpha_4$. As a control, we also synthesized H2$\alpha_2$, which consists of $\alpha_2$ with two histidine residues substituted on a single helix. If the peptides containing three histidine residues coordinate $Zn^{2+}$ using all three histidine residues, the control peptide should bind $Zn^{2+}$ with reduced affinity. Plate 8 shows a molecular model of one subunit of H3$\alpha_2$ and figure 3.8 shows the four sequences.

All three proteins assemble into 4-helix bundle proteins similar to $\alpha_2$ and $\alpha_4$. NMR spectroscopy indicates that H3$\alpha_2$ binds $Zn^{2+}$ in a 1:1 complex. The resonances associated with the $\delta$ and $\epsilon$ protons of the three histidine residues of H3$\alpha_2$ are broad and overlapped in the absence of $Zn^{2+}$. However, upon addition of a stoichiometric amount of $Zn^{2+}$, six sharp and well-resolved resonances are observed (figure 3.9a). This indicates that H3$\alpha_2$ forms a well-defined metal complex in which exchange between free and bound $Zn^{2+}$ is slow on the NMR time scale. It seems that all three histidine residues ligate the metal because all histidine resonances show substantial changes in chemical shift upon

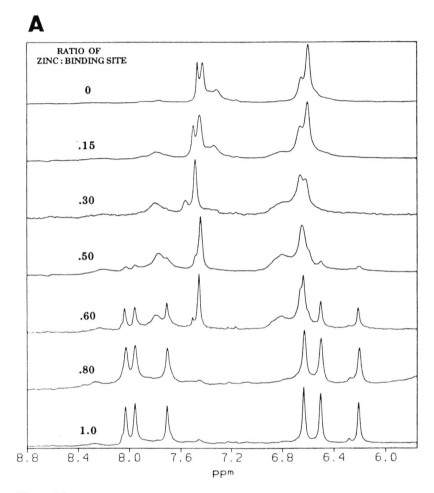

**Figure 3.9**
Resolution of NMR resonances of H3$\alpha_2$ with the addition of Zn$^{+2}$ (A). Resolution of NMR resonances of H2$\alpha_2$ with the addition of Zn$^{+2}$ (B).

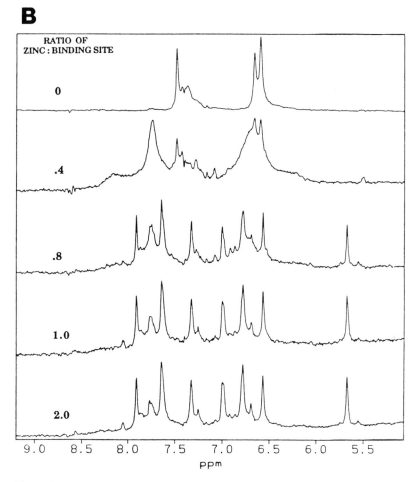

Figure 3.9
(Continued.)

addition of zinc. Titration of H2$\alpha_2$ with Zn$^{+2}$ also indicated that Zn$^{+2}$ was coordinated with the ligands. However, above stoichiometric concentrations of Zn$^{+2}$, multiple resonances indicate that a unique complex is not formed (figure 3.9b).

The H3$\alpha_4$ protein also formed a 1:1 complex with Zn$^{+2}$. In contrast with H3$\alpha_2$ and H2$\alpha_2$, this protein has only a single metal site per bundle. The NMR spectra of H3$\alpha_4$, although altered by the addition of Zn$^{+2}$, remains broad even above stoichiometric amounts of metal (figure 3.10a) and is reminiscent of the spectral features of H3$\alpha_2$ at stoichiometries of Zn$^{+2}$/binding site below 0.5 (that is, when only one Zn$^{+2}$ is bound per bundle). However, with H6$\alpha_4$, which has two sites per bundle, the proton resonances of the six histidine residues are sharp and well defined (figure 3.10b). Thus it appears that two occupied sites create a more structured and less dynamic bundle than bundles with only one occupied site or those without any metal ions at all. This illustrates the additive nature of specific interactions in structuring these bundles.

The conformational stability of the peptides in the presence and absence of Zn$^{+2}$ was assessed by circular dichroism. Figure 3.11 illustrates the dependence of $\theta_{222}$ for H3$\alpha_2$ and H3$\alpha_4$ as a function of guanidine-HCl concentration. In the absence of Zn$^{2+}$, H3$\alpha_2$ is less stable than $\alpha_2$ (data not shown). This is expected because apolar leucine residues have been substituted for polar histidine residues at (relatively buried) positions thought to be important for helix-helix packing. In the presence of 1.0 mM Zn$^{2+}$, the stability of H3$\alpha_2$ increases and the free energy of dimerization becomes more favorable by 1.9 kcal/(mol bound Zn$^{2+}$). Likewise, H3$\alpha_4$ unfolds at considerably higher guanidine concentrations and showed an increase in stability of 2.8 kcal/mol in the presence of 1.0 mM Zn$^{2+}$. In contrast, $\alpha_2$ shows no change in stability in the presence of Zn$^{2+}$. This provides further evidence that the three histidine residues coordinate the metal.

## Conclusion

To summarize, I have tried to describe the approach and progress in the DeGrado group in the design and characterization of 4-helix bundle proteins. These data demonstrate that we have engineered proteins with desired helical secondary structural features. NMR evidence indicates

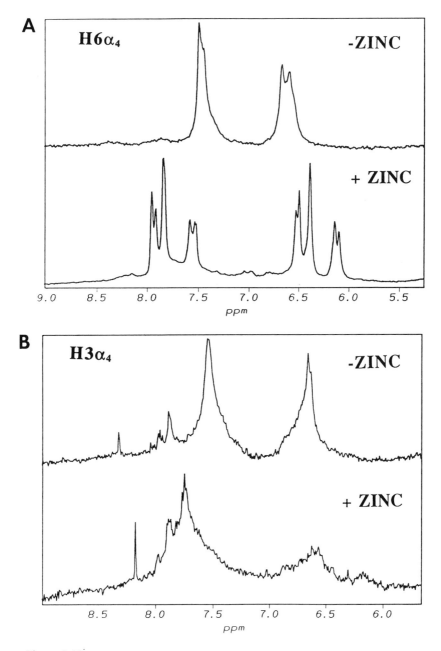

**Figure 3.10**
Proton spectra of H3α₄ (A) and H6α₄ (B) in the presence and absence of zinc.
Samples were prepared in a buffered solution containing D₂O so that only
histidine resonances are apparent.

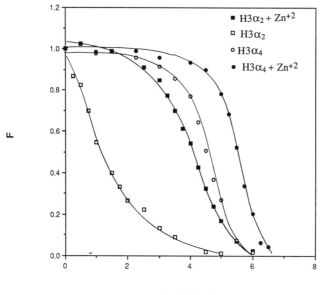

**Figure 3.11**
Variations of $\theta_{222}$ for H3$\alpha_2$, and H3$\alpha_4$ as a function of guanidine-HCl
concentration.

that the tertiary structure is at least consistent with some elements of
the design. What is not yet clear, however, is the uniqueness of the
tertiary packing nor the bundle's dynamic quality. These bundle proteins
may have some characteristics intermediate between "molten globule"
and native protein states. In contrast to the native state, which has well-
defined secondary and tertiary structures, the molten globule state is
best described as pH-, temperature-, or denaturant-dependant. The tertiary structure of molten globule proteins is loosely organized and/or
fluctuates due to the reduction of specific tertiary contacts.

While the extraordinary stability of these designed bundles indicate
that tertiary contacts are in abundance, we are currently questioning the
uniqueness of these contacts. It is reasonable to imagine that the hydrophobic core, which consists essentially of a nonspecific hydrophobic
surface, could be thermodynamically satisfied by a large number of
structural variations in helical packing. To explore this question, we
have introduced specific interactions between the helices, such as metal-

binding sites. Increased dispersion in the NMR spectra, as well as the induction of temperature-dependent unfolding transitions, indicate that metal sites do indeed confer more nativelike character to the 4-helix bundles. Ultimately, we would like to design and build catalytically active proteins. Metal-binding 4-helix bundles represent a significant step toward this goal.

## Notes

I would like to recognize those who have contributed to various aspects of this work. In particular, Bill DeGrado has been responsible for the genesis of this and many other protein design efforts. Acknowledgments and thanks to: Peter Connally, William DeGrado, Peter Domaille, David Eisenberg, Siew Ho, Jeff Hoch, David Live, John Osterhout, Lynne Regan, and Zelda Wasserman.

1. L. Regan, W. DeGrado, "Characterization of a helical protein designed from first principles," *Science* 241 (1988): 976–978.

2. S. Ho, W. DeGrado, "Design of a 4-helix bundle protein: Synthesis of peptides which self-associate into a helical protein," *J. Amer. Chem Soc.* 109 (1987): 6751–6758.

W. DeGrado, L. Regan, S. Ho, "The design of a four-helix bundle protein," *Cold Spring Harbor Symposia on Quantitative Biology* 2 (1987): 521–526.

3. The α-helix is one of the basic elements of protein secondary structure into which a polypeptide (protein) chain can fold. (The linear sequence of amino acid residues that are linked sequentially to form the protein is referred to as the primary structure of the protein, while the complete three-dimensional structure into which the protein folds, including the further folding of the elements of secondary structure, is referred to as the tertiary structure.) The α-helix is a rodlike structure in which the polypeptide chain is tightly coiled with the polymer backbone inside the helix and with the amino acid side chains exposed to the outside. The helix is stabilized by hydrogen bonds between the backbone NH and CO groups on amino acid residues separated by four positions in the sequence. See L. Stryer, *Biochemistry* (New York: W. H. Freeman and Company, 1988).

4. In most proteins that contain α-helices, the α-helix is present as only a short rod (usually less than 40 Å in length). In other proteins, two or more helices as long as 1000 Å can intertwine in a superhelical structure to form α-helical coiled coils. This motif is often found in structural proteins such as keratin in hair, myosin in muscle, and fibrin in blood clots.

5. See E. O'Shea, R. Rutkowski, P. Kim, "Evidence that the leucine zipper is a coiled coil," *Science* 243 (1989): 538–542.

See also W. Landshultz, P. Johnson, and S. McKnight. "The leucine zipper: A hypothetical structure common to a new class of DNA binding proteins," *Science* 240 (1988): 1759–1764. It seems that the ability of this motif to promote dimerization of DNA-binding gene regulation factors makes possible the combinatorial diversity of gene regulation factors through the dimerization of either identical molecules or of molecules of the same family of factors.

6. See J. Lear, Z. Wasserman, W. DeGrado, "Synthetic amphiphilic peptide models for protein ion channels," *Science* 240 (1988): 1177–1181.

7. See W. DeGrado, "Design of peptides and proteins," *Advances in Protein Chemistry*, vol. 39, ed. C. B. Anfinsen, J. D. Edsall, F. M. Richards, D. S. Eisenberg (New York: Academic Press, 1988).

Also see W. DeGrado, Z. Wasserman, J. Lear, "Protein design, a minimalist approach," *Science* 243 (1989): 622–628.

8. H. Scheraga, "Use of random copolymers to determine the helix-coil stability constants of the naturally occuring amino acids," *Pure and Appl. Chem.* 50 (1978): 315.

P. Lui, M. Liff, L. Marky, N. Kallenbach, "Side chain contributions to the stabilities of alpha-helical structure in peptides," *Science* 250 (November 1990): 669–673.

K. O'Neill, W. DeGrado, "A thermodynamic scale for the helix-forming tendencies of the commonly occurring amino acids," *Science* 250 (1990): 669–673.

Baldwin, "Relative helix-forming tendencies of nonpolar amino acids," *Nature* 344 (March 1990): 268–270.

9. Two approaches have been used to determine the preferences of individual amino acids for being in an α-helix. Both focus on the influence of short-range interactions and ignore the importance of longer range interactions in stabilizing secondary structure. The simplest approach is statistical (exemplified by the Chou Fasman method) in which the frequency of occurrence of a given amino acid in helical versus nonhelical structures is determined by analysis of known protein crystal structures.

An alternative experimental approach is the host-guest method introduced by Scheraga. The extent to which a given amino acid stabilizes or destabilizes the helical conformation is determined by examining amino acid copolymers. The amino acid of interest is inserted as a "guest" into a random copolymer made with a large molar excess of "host" amino acid, which is chosen because it favors helix formation. The transition between helix and random coil is measured for the copolymer and for a homopolymer of the host amino acid alone, and the contribution of the guest to helical stability is estimated from a mathematical model of helix formation.

10. K. Shoemaker, P. Kim, E. York, J. Stewart, R. Baldwin, "Tests of the helix dipole model for stabilization of α-helices," *Nature* 326 (1988): 563–567.

The term "helix macrodipole" refers to the fact that in an a-helix, the (positively charged) amide NH groups point approximately toward the N-terminus of the helix while the (negatively charged) $C = O$ groups point in the opposite direction, forming a macrodipole (a separation of charge along the length of the helix). Thus negatively charged residues at the N-terminus or positively charged residues at the C-terminus interact favorably with the helix dipole and therefore promote helix formation.

11. P. Weber, F. Salemme, "Structural and functional diversity in 4 α-helical proteins," *Nature* 287 (1988): 82–83. The 2,2,2 symmetry is an idealization of the symmetry found in the structures of natural 4-helix bundle proteins.

12. T. Creighton, *Proteins: Structure and Molecular Principles* (New York: W. H. Freeman and Company, 1984), 179–182. All amino acids except glycine have an asymmetric carbon. In addition, peptides constructed from the amino acids may adopt a conformation such as a helix which is either right or left

handed. The resulting intrinsic asymmetry of peptides causes them to absorb differentially right- and left-circularly polarized light that can be measured by circular dichroism (CD).

13. J. Osterhout, T. Handel, G. Na, A. Toumadje, R. Long, P. Connolly, J. Hoch, W. Johnson, D. Live, W. DeGrado, "Characterization of the structural properties of $\alpha_1 B$, a peptide designed to form a 4-helix bundle," *J. Am. Chem. Soc.* 114 (1992): 331–337.

14. See J. Markley, "One and two-dimensional NMR spectroscopic investigations of the consequences of amino acid replacements in proteins," *Protein Engineering,* ed. D. Oxendar, C. Fox (New York: Alan L. Liss, Inc., 1987) 15–33 for a brief review of the use of nuclear magnetic resonance (NMR) in the determination of protein structure. For an extended discussion, see K. Wüthrich, *NMR of Proteins and Nucleic Acids,* (New York: John Wiley & Sons, 1986).

15. S. Englander, N. Kallenbach, "Hydrogen exchange and structural dynamics of proteins and nucleic acids,"*Q. Reviews Biophysics* 16 (1984): 521–655.

16. Because there is an equilibrium between $\alpha_1 B$ in the tetrameric (folded) state and the monomeric (unfolded) state, the exchange is concentration dependent. As the concentration is decreased and the monomeric species becomes populated, the exchange rates become faster. The monomeric state is relatively unfolded and can exchange protons for deuterons more readily than in the tetrameric state where the amide protons are more protected by hydrogen bonding.

See R. Molday, S. Englander, R. Kallen, "Primary structural effects on peptide group exchange," *Biochemistry* 11 (1972): 150–158.

17. The protection factors were calculated from the exchange rates of $\alpha_1 B$ at a concentration where $\alpha_1 B$ is predominantly (> 99%) tetramer.

18. The exception to this is glycine, which can have two cross-peaks due to its two $C_\alpha \underline{H}$ proton, and proline, which has no $C_\alpha \underline{H}$ protons and therefore does not show a "fingerprint."

19. See G. Vuister, R. Boelens, R. Kaptein, "Non selective three-dimensional NMR spectroscopy. The 3D HOHAHA experiment," *J. Mag. Res.* 80 (1988): 176–185.

See also H. Oshkinat, C. Cieslar, C. Griesinger, "Recognition of secondary-structure elements in 3D TOCSY-NOESY spectra of proteins. Interpretation of 3D cross-peak amplitudes," *J. Mag, Res.* 86 (1990): 453–469.

20. See K. Kuwajima, "The molten globule state as a clue for understanding the folding and cooperativity of globular-protein structure," *Proteins: Structure, Function and Genetics* 6 (1989): 87–104.

21. T. Handel, W. DeGrado, "De novo design of a $Zn^{2+}$-binding protein," *J. Am. Chem. Soc.* 112 (1990): 6710–6711.

22. C. Noren, S. Anthony-Cahill, M. Griffith, P. Schultz, "A general method for site specific incorporation of unnatural amino acids into proteins," *Science* 244 (1989): 182–188.

See also, J. Bain, C. Glabe, T. Dix, A. Chamberlin, "Biosynthetic incorporation of a non-natural amino acid into a polypeptide," *J. Amer. Chem. Soc.* 111 (1989): 8013–8014.

23. T. Sasaki, E. T. Kaiser, "Helichrome: synthesis and enzymatic activity of a designed hemeprotein," *J. Am. Chem. Soc.* 111 (1989): 380–381.

24. See D. Eisenberg, W. Wilcox, S. Eshita, P. Pryciak, S. Ho, W. DeGrado,

"The design, synthesis, and crystallization of an alpha-helical peptide," *Proteins: Structure Function and Genetics* 1 (1986): 16–22.

See also C. Hill, D. Anderson, L. Wesson, W. DeGrado, D. Eisenberg, "Crystal structure of $\alpha_1$: implications for protein design," *Science* 243 (August 1990): 543–546.

25. D. Ciesla, D. Gilbert, J. Feigon, "Secondary structure of the designed peptide alpha-1 determined by nuclear magnetic resonance spectroscopy," *J. Am. Chem. Soc.* 113 (1991): 3957–3961.

26. See K. O'Neil, R. Hoess, W. DeGrado, "Design of DNA-binding peptides based on the leucine zipper motif," *Science* 249 (1990): 774–778.

27. C. Murre, P. McCaw, H. Vaessin, M. Caudy, L. Jan, Y. Jan, C. Cabrera, J. Buskin, S. Hauschka, A. Lassar, H. Weintraub, D. Baltimore, "Interactions between heterologous helix-loop-helix proteins generate complexes that bind specifically to a common DNA sequence," *Cell* 56 (1989): 777.

## Discussion

*Ward:*   When the zinc is incorporated into the bundle, is it incorporated into a preformed bundle of the proteins or do the proteins assemble around it?

*Handel:*   The metal sites are on the interface between two helices and on the solvent-exposed side of the bundle. Therefore, the site is accessible to metal even if the bundle is preformed. In the absence of metal, the peptides dimerize to give a 4-helix bundle, so, in principle, the metal could bind to a preformed site. At concentrations of peptide where we have mostly 4-helix bundles, it is likely that this is the case. However, the thermodynamic stability of the bundle in the absence of metal is lower than in the presence of metal. Consequently, at concentrations of peptide where we would have a significant concentration of dimer in equilibrium with tetramer, if we add metal (which increases the association constant $k_a$), the metal promotes tetramer formation.

When the metal binds to a preformed site (i.e., at concentrations where we have the tetramer even in the absence of metal), we have NMR and calorimetric evidence that the metal further structures the protein.

*Audience:*   Are you trying to stay with natural amino acids as chelators so that you can clone all of these proteins?

*Handel:*   For the moment, yes. From the standpoint of trying to address questions relating to protein folding, using the DeGrado "minimalist" approach of generating desired folding motifs from the simplest amino acid sequence, we are content to use natural amino acids. Furthermore, from the standpoint of following up our designs with NMR structure characterization, we are also interested in expressing proteins in E. coli so we can generate large quantities of protein with isotopic labels (i.e., $^{15}N$ and $^{13}C$). Biosynthesis is currently routine only with natural amino acids.

On the other hand, since we do make many of our peptides synthetically, we could introduce nonnatural amino acids as, for example, good ligands for metals or as participants in catalytic reactions. Although nonroutine, a few groups, such as Peter Schultz's,[22] have made great progress in the biosynthetic incorporation of unusual amino acids and this is likely to become more feasible in the future.

*Audience:* For instance, if you were trying to bind zinc, could you use a porphyrin or something else, rather than histidines?

*Handel:*   Yes. In fact this has already been done by Kaiser and Sasaki.[23] In this work, porphyrin was used as a template to which four amphipathic helices were attached; some hydroxylase activity was even observed.

*Audience:*   What are these proteins going to be used for?

*Handel:*   Right now the application we have in mind for the 4-helix bundles is for making catalysts. On the other hand, there are many naturally occurring DNA-binding proteins with purported helix-loop-helix folding motifs, and one can think about making bundles which recognize and regulate DNA. In the case of the ion channel project, all kinds of biosensors are possible. For example, we could attach receptors to the surface of the ion channel in such a way that when the receptor recognizes its ligand, it blocks the channel or disaggregates the helices, thereby effectively leading to "channel closing." Our main interest right now, however, is the fundamental one of understanding the protein folding problem. This is the greatest challenge.

*Audience:*   Have you tried different ways of breaking the structure? In particular, have you tried varying the charges or putting small hydrophobic residues in the cavity. Do you have any idea how robust it is in that sense?

*Handel:*   We have not tried to disrupt the structure on purpose. However, in the course of generating mutants of $\alpha_1B$, we have been able to generate several forms that are slightly destabilized and some that are completely unfolded. If you substitute alanine and threonine for $Leu_3$ and $Leu_{10}$ in $\alpha_1B$, respectively, the peptide has minimal helical content and does not tetramerize. Likewise, our first attempt at a metal-binding protein was to replace $Leu_6$ with a histidine in order to get a tetrahedral zinc site at the interior of the bundle, but that substitution greatly destabilized the bundle. (Histidine is a relatively helix-destabilizing amino acid as is threonine.) We also made single mutants which have phenylalanine in place of $Leu_6$ or $Leu_{10}$. The $Leu_6Phe$ mutant was destabilized by only $\sim 1$ kcal/mole, whereas $Leu_{10}Phe$ was destabilized by $\sim 4$ kcal/mole. If one examines the computer models of $\alpha_1B$, there is a hole in the vicinity of $Leu_6$ which might be filled partially by a phenylalanine substitution, whereas the packing around $Leu_{10}$ is tighter and there may be less room to accommodate a phenylalanine.

How robust is it? I think that very much depends on the nature of the substitution (as demonstrated by the difference in the effects of a $Phe_6$ vs $His_6$ substitution). Since $\alpha_1B$ is only a 16-residue peptide, one might expect that certain single mutations could greatly destabilize the structure since they are amplified by four in the bundle.

*Audience:*   Has your protein been expressed in E. coli?

*Handel:*   Yes, $\alpha_4$ has been expressed in E. coli by Lynne Regan (see reference 1). It has also been synthesized by solid-phase methods, but not without difficulty. For a protein of this size, it is worthwhile to express rather than synthesize it as the expression also facilitates generating mutants.

*Audience:*   What about David Eisenberg's crystal structure?

*Handel:*   A crystal structure of $\alpha_1$ has been solved by Chris Hill and David Eisenberg.[24] This is a 12-residue peptide which was a by-product of the synthesis of $\alpha_1B$ with the structure Acetyl-Glu-Leu-Leu-Lys-Lys-Leu-Leu-Glu-Glu-Leu-Lys-Gly-COOH. The crystal structure is more complex than a simple tetramer—it contains both tetramer and hexamer. It's a very interesting structure. The fact that it is not a 4-helix bundle is not surprising, however, because the peptide is much less stable than $\alpha_1B$, and the peptide was crystallized at low pH which is destabilizing.

Since it is relatively unstable, it is likely that crystal packing forces have an affect on the structure. The solution conditions also seem to have an affect since a sulfate ion is involved in the crystal structure and since gel filtration and NMR studies of the peptide at higher pH indicate that the crystal and solution structures are not the same.[25]

*Audience:* What is the status of crystallography under the conditions that you have used in your work?

*Handel:* We collaborate with David Eisenberg's group and they are trying to get crystals of various bundle peptides. In particular we have ligated our metal-binding peptide H3$\alpha_2$ with $Co^{+2}$ and oxidized the $Co^{+2}$ to $Co^{+3}$ which is a slowly dissociating ligand. Hopefully that will help the crystallization.

*Drexler:* The $\alpha_4$ is currently my favorite protein, so I'm naturally concerned about it. Clearly it would be more stable in forming a structure if there were multiple packings in the core that were all accessible because then there would be entropic factors favoring one or more of these folded states, but then it would not be a specific folded state. Do you have either experimental or theoretical grounds for excluding these alternatives?

*Handel:* We do not have evidence to exclude multiple conformations. In fact multiple conformations could be the basis of the symmetry that we see in the NMR spectrum. Furthermore, as you mentioned, if the protein adopts multiple conformations, this would be energetically favorable because the entropy loss accompanying folding would not be as great as for a protein with a unique conformation. We are planning to address these questions with a variety of techniques including NMR relaxation experiments which give dynamic information. We also plan to introduce tryptophan into $\alpha_4$ so we can carry out time-resolved fluorescence anisotropy decay measurements to probe the time scale of side chain fluctuations.

*Audience:* Could you make DNA-binding proteins with these bundles?

*Handel:* That's a very exciting thought! Karyn O'Neil, Ron Hoess, and Bill DeGrado have designed DNA-binding proteins based on the leucine zipper motif.[26] As I mentioned before, we could try and make DNA-binding proteins based on the helix-loop-helix motif.[27]

*Audience:*    What sort of catalytic activity are you thinking of looking at in the future?

*Handel:*    To start with, simple hydrolase, nuclease, or protease activity, ultimately with specificity. Also, we are interested in making metal sites with electron transfer potential.

# 4

## Design of Self-Assembling Molecular Systems: Electrostatic Structural Enforcement in Low-Dimensional Molecular Solids

Michael D. Ward

This chapter outlines some of our strategies to design and manufacture molecular materials based on their interatomic and intermolecular interactions—building materials "from the bottom up." The chapter describes:

- The concept of molecular self-assembly
- Components currently available for solid state organic systems
- Interatomic interactions in self-assembling crystalline materials
- Design techniques for engineering organometallic materials
- Using electrostatic templates to control dimensionality
- Building structures from small molecular assemblies

This work was initiated in order to construct macroscopic electronic devices. At first we focused on functional materials that exhibited a measurable property in order to determine how the structure of that material influenced a particular electronic function. In this chapter, however, the electronic aspects of these materials are presented only briefly in order to emphasize instead structural properties.[1]

### Molecular Self-Assembly

Considering crystallization in terms of self-assembly (figure 4.1) has a number of advantages. X-ray crystallography has provided a great deal of structural information on individual molecules: bond lengths, bond angles, positions of atoms, and so on. Recent efforts, however, have focused on the intermolecular interactions that determine the structure of a three-dimensional solid.

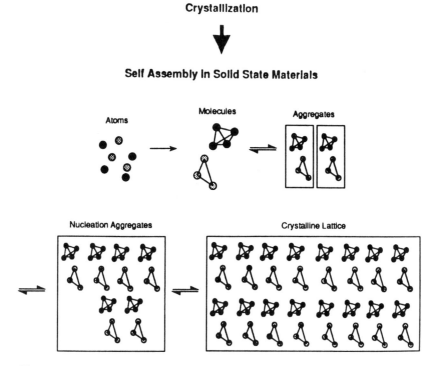

**Figure 4.1**
Crystallization as self-assembly of solid-state materials.

Describing crystallization as self-assembly addresses the relationship between the position of molecules within a crystalline lattice and the aggregates that are responsible for the evolution of that lattice. Inspection of a known crystal structure can allow one to infer the structure of the small molecular aggregates that assemble to form that crystal, and, conversely, rational design of molecular aggregates can lead to a molecular solid with particular functional properties.

It is important to understand the process of self-assembly in the solid state in order to control the structural architecture of solids. Intensive study of solid-state organic chemistry began in the 1970s. Gerhard Schmidt coined the term "crystal engineering" during his pioneering studies of topochemical reactions in solids and their relationship to the disposition of matter in those solids.[2,3] *Topochemical reactions* are reactions in which two distinct molecules in the solid state can be induced to react to form dimers and other polymers. Synthetically, this is not a very high yield process, but it is interesting from the point of view of making materials and also from the point of view of information storage.

Understanding self-assembly in the solid state can also have a number of practical consequences. For one, similar architectural principals may also be important in molecular recognition processes. To date, however, the study of topochemical reactions has focused on attempts to control the structural architecture of a solid in order to create materials with desirable magnetic, electronic, or optical properties such as nonlinear effects. An understanding of the growth and dissolution of crystals, and how to control crystal morphology in the crystallization process, has also had great impact in the pharmaceutical area.

### Interactions in Self-Assembly of Crystalline Materials

The key factors in building solids "from the bottom up" are their interatomic and intermolecular interactions. The forces included in self-assembling crystalline systems are similar to those in proteins and include electrostatic, van der Waal's, $\pi$-$\pi$ interactions, dipole attractions, hydrogen bonding, and electron delocalization. A key principal in crystalline materials is the tendency for crystals to minimize void space, which is a direct result of the maximization of attractive forces.

Figure 4.2 shows six of the many self-assembled systems that are held together primarily by electrostatic interactions. Electrostatic interactions account for a great deal of energy in crystalline structures. This chapter presents some nontraditional materials that take advantage of electrostatic interactions but are unlike these solids in that they are not represented by point charges in a lattice.

In the materials described here, charge transfer interactions are also important. We have been interested primarily in charge transfer interactions because they provide a strategy for the design of molecular solids as well as desirable electronic, magnetic, and optical properties. Probably the most notable example of this class of compounds is tetrathiafulvalene-tetracyanoquinodimethane (TTF-TCNQ), shown in figure 4.3.

TTF-TCNQ, discovered in the 1960s, was the first proven organic conductor. It conducts by virtue of overlapping $\pi$-electron systems, which, when stacked in extended chains, form a metallic band capable

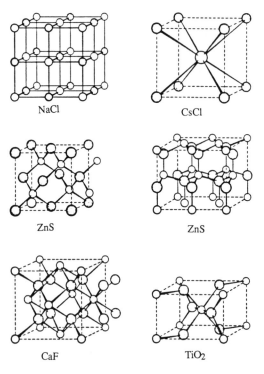

NaCl

CsCl

ZnS

ZnS

CaF

TiO$_2$

**Figure 4.2**
Electrostatic interactions in crystalline matrices.

of supporting electron mobility through the stack. This particular "segregated stack" structure is important for the formation of metallic bands in numerous organic solids commonly resulting in either semiconducting or metallic properties.[4]

TTF-chloranil displays a very different structure from the segregated stack structure of TTF-TCNQ (figure 4.3). (Chloranil is a chlorine substituted quinone: $C_6Cl_4O_2$.) TTF-chloranil is a "mixed stack" structure which contains alternating TTF and chloranil molecules. They are generally not conducting, but they do have interesting optical properties resulting from charge transfer interactions, in this case between TTF and chloranil. These two types of structures comprise the two major classes of materials known as "low dimensional solids."[5]

Hydrogen bonded solids have recently attracted greater attention. Organic chemists have long been familiar with hydrogen bonding, but studies of crystal structures have revealed that hydrogen bonds may be very important for directing molecular assembly in the solid state. Resorcinol has several different types of hydrogen bonds, indicated by the dashed lines in figure 4.4. Nitroanilines, which have been studied for their nonlinear optical properties, likewise exhibit significant hydrogen bonding interactions.

Although an individual hydrogen bond is not energetically very significant, summed over an entire crystal they can have a large effect on crystal structure. Adamantane tetracarboxylic acid has a unique structure that could be described as five interpenetrating diamond lattices (figure 4.4c).[6] Unfortunately, a two-dimensional figure does not do this structure justice.

Halogen steering effects have also been observed.[7] Chlorine atoms (shown as large circles in figure 4.5a) are capable of "steering" structures of dimers to this beta structure.[8] It seems that chlorine-chlorine interactions are very important in determining structures in solids.

Figure 4.5b displays another significant weak interaction, the CH-O interaction. The structure of 3,4-methylenedioxycinnamic acid was found to be *syn*planar (rather than *anti*planar). This conformational preference was attributed to a C-H . . . O bond of 2.48 Å between the alpha-olefinic hydrogen and a heterocyclic oxygen. This is an example of the type of weak hydrogen bonding interactions that have become more appreciated recently, and it is important in determining the structure of this cinnamic acid derivative.[9]

Illustrated in figure 4.6 is an example of the complexity encountered in designing a molecular solid: bis(ethylenedithio)tetrathiafulvalene (called ET for short). Materials containing this molecule have been thoroughly studied by Jack Williams and his collaborators at Argonne Laboratories. Not only does ET have numerous polymorphs and forms numerous different compounds with different anions (X in figure 4.6), but the structure also shows the richness of molecular interactions described above.

Furthermore, all of these interactions seem to be important in determining the properties of ET. The rings are flat and can thus approach each other closely. This is desirable for close packing of molecules to minimize void space in the material. There are hydrogen bonding inter-

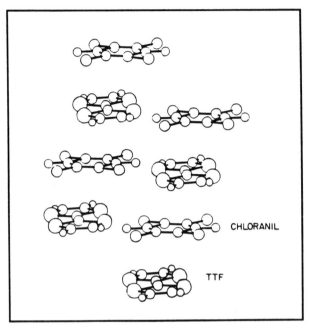

TTF-TCNQ
Segregated stack

**Figure 4.3**
Charge transfer interactions in crystalline matrices.

actions between the anions and hydrogen atoms on the carbon skeleton. Sulfur-sulfur interactions are responsible for the three-dimensional nature of this material and are very important in determining the electronic properties (ET is a superconductor at low temperatures). The $\pi$-$\pi$ interactions are important for forming metallic bands, and these bands help to drive the formation of the material by providing an increase in overall stability.

ET is also a good example of molecular engineering of a material. You could not easily predict this structure ab initio, but once it is known that the cation tends to form horizontal sheets, which in turn stack vertically, the size of the anion can be systematically changed. That has been done very nicely by changing the size of the linear anions—triiodide, gold iodide, and dibromo iodide; the decrease in anion size exerts

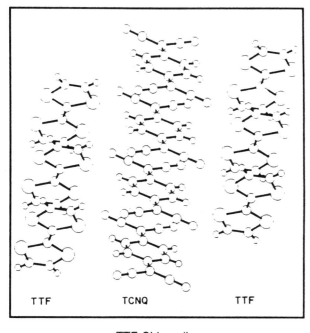

TTF             TCNQ            TTF

TTF-Chloranil
...DADADA...

**Figure 4.3**
(Continued.)

a lattice pressure and squeezes the molecules together. The shorter anions suppress the transition to a semiconducting or to a metallic state and actually favor the superconducting transition; the onset of superconductivity occurs at a higher temperature. A common theme that emerges with these materials is that *small* systematic changes in the solid state can afford desirable properties in a predictable fashion.

The significance of $\pi$-$\pi$ interactions and electron delocalization may be greater in low-dimensional solids, but in general the concepts of close packing to minimize void space, perpendicular stacking of aromatic residues, and so on, are very important in both crystalline systems, much like proteins. Inspection of crystallographic data bases to determine the amount of void space in solid materials reveals that 65% to 77% of the total volume of a crystal is occupied by atoms. This represents a balance

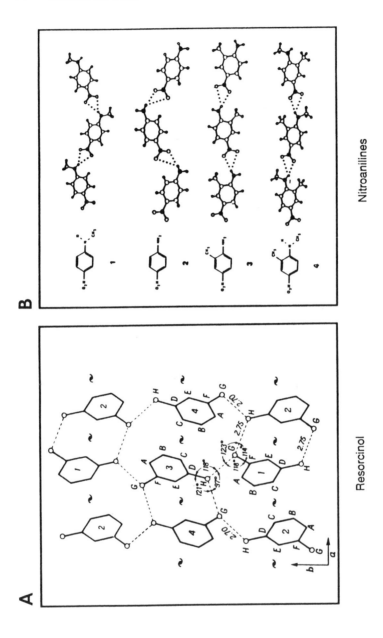

Resorcinol          Nitroanilines

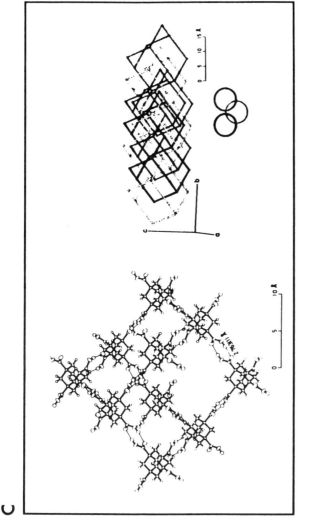

**Adamantane-1,3,5,7-tetracarboxylic acid**

**Figure 4.4**
Hydrogen bonding in crystalline matrices. (Figure 4.4a reprinted with permission from J. M. Robertson, *Proc. of the Royal Society A*, 157 (1936): 133. Copyright 1936 The Royal Society. Figure 4.4b reprinted with permission from T. W. Panunto, Z. Urbanezyk-Lipkowska, R. Johnson, N. C. Etter, *J. Am. Chem. Soc.* 109 (1987): 7786–7797. Copyright 1987 American Chemical Society. Figure 4.4c reprinted with permission from O. Ermer, *J. Am. Chem. Soc.* 110 (1988): 3747. Copyright 1988 American Chemical Society.)

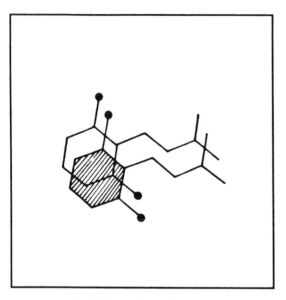

2,6-dichlorocinnamic acid

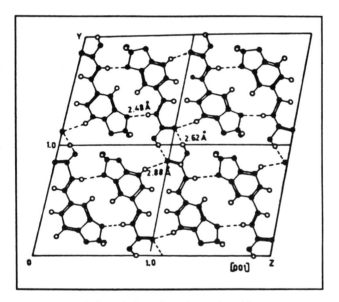

3,4-methylenedioxycinnamic acid

**Figure 4.5**
Halogen "steering" effects in crystalline matrices (above). C-H . . . O interactions in crystalline matrices (below).

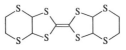

β-(ET)₂X

☐ Hydrogen-bonding Interactions

☐ Electrostatic Interactions

☐ S-S Interactions

☐ π–π Interactions

☐ Metallic Band Formation

| X | $I_3^-$ | $AuI_2^-$ | $IBr_2^-$ |
|---|---|---|---|
| length (Å) | 9.9 | 9.4 | 9.3 |
| $T_c$ | 1.4 | 5.0 | 2.8 |

$I_3^-$ too large to fit into hydrogen pocket ⇨ structural modulation

**Figure 4.6**
Bis-ethylene-dithio-tetrathiafulvalene (ET). (Courtesy J. Williams, K. Carneiro. 1985. *Inorg. Chem. Radiochem.* 29: 249.)

of repulsion terms and dispersion forces that allow molecules to get close enough together to pack the solid state.

## Design Techniques

In the early days of molecular modeling, mechanical devices that manipulated hard-sphere models of molecules were used to determine the minimum volume of a crystal structure. Today, computer-aided design techniques provide more efficient means for predicting structures.

The first part of this approach involves the extensive X-ray crystallography data base that describes atom connectivities and the effects of atomic substitutions and functional group substitutions on molecules. The Cambridge crystal data base lists over 80,000 structures, and this is only a small fraction of the number of known compounds. Interest in crystallographic data bases will grow as people strive to understand packing arrangements.

❑ Experience with previous structures

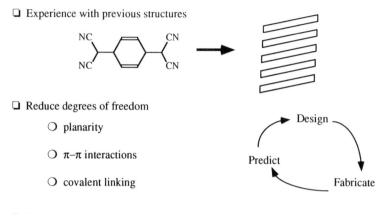

❑ Reduce degrees of freedom

    ○ planarity

    ○ π–π interactions

    ○ covalent linking

❑ Best guess

    Reality ⇨ Currently difficult to predict

    Result ⇨ Can predict structures within a narrow class

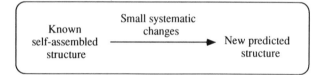

**Figure 4.7**
Intuitive techniques in designing molecular solids.

Molecular modeling is used to complement known X-ray structures to predict the packing of molecules in the solid state. Molecular modeling software is used to minimize the energy of connected structures and to build structures that minimize void space by balancing repulsion and dispersion forces. Electrostatic Madelung energies can be taken into account when calculating structures that are charged or have many polar functions.[10]

Using these approaches, structures can be designed and some predictions about their properties can be made. In actuality, however, we still use chemical intuition. We work with systems containing molecular components with which we have had a lot of experience (figure 4.7). TCNQ is ubiquitous in its appearance in low-dimensional solids. From past experience we know that TCNQ—in its anionic form—tends to stack in one-dimensional columns in which the planes overlap. We use molecules with known patterns of intermolecular interactions, such as TCNQ, because they constrain the solid-state lattice to a smaller number of possible structures. In almost all cases it will be forced to line up in a single stack. We thereby take advantage of molecular properties, including planarity, $\pi$-$\pi$ interactions, and, in some examples, covalent linking between molecules.

To develop a new material, we take the following steps:

1. Predict a structure.
2. Design a system.
3. Fabricate the system.
4. Study the structure.
5. Redesign with systematic changes to optimize structure.

Eventually these steps converge to where the prediction is borne out by the experiment. In general, we can not yet do this very easily. However, the task is simplified if we start with known structures, then systematically vary a small part and try to predict the new structure. It is especially important to attack systems where we can reduce the number of degrees of freedom available to the components.

P. J. Fagan and I thought to take advantage of Madelung energy in solids by devising a system of compounds that had various degrees of charge that were spatially oriented in different fashions. For example, imagine a point charge as a zero-dimensional cation. Then two cations linked together would be a one-dimensional rod of two positive charges. If you can make triangular, square, pentagonal, and hexagonal spatial arrangements of charge, can the same type of structural topology be conferred in the solid?

Planar anions (TCNQ and related compounds) tend to stack in a one-dimensional sense. Some of the motifs that have been observed are depicted in figure 4.8: cations sandwiching two anions in which the anions interact electronically, mixed stacks such as TTF-chloranil, and segregated stacks such as TTF-TCNQ (figure 4.5). But, if you can form segregated stacks, what happens if the charges are moved farther apart? Is it possible for neutral molecules to insert between the anions and change the electronic properties of that system?

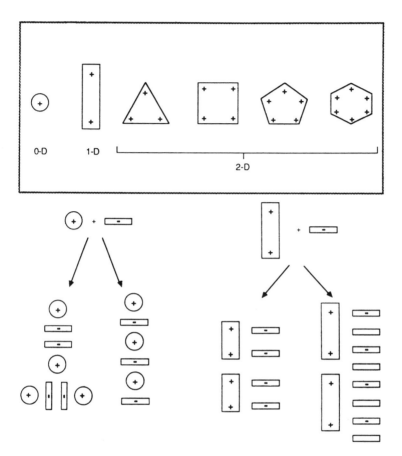

**Figure 4.8**
Schemes for conferring structure via the spatial arrangement of charge.

## Molecular Solids with Organometallic Components: Structure and Function Relationships

To study "zero"-dimensional cations, we first looked at some highly charged organometallic complexes called sandwich complexes (figure 4.9). These complexes have a flat ring structure. These materials are also redox active; that is, they can donate or accept electronic charge and thus might have interesting charge transfer properties. Consequently we thought to learn something about structure-function relationships in these materials. One can think of such an arene-iron or ruthenium

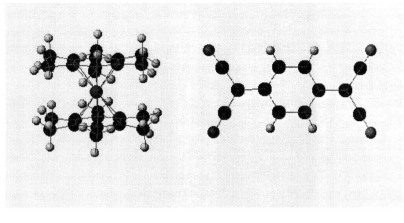

**Figure 4.9**
Structures of sandwich complex (point charge) dication $[(\eta^6\text{-}C_6Me_6)_2M]^{2+}$ and TCNQ anion.

dication as a "fat" benzene molecule.[11] The distance between rings is about 3.5 Å, depending on the identity of the metal. We hoped that planar TCNQ anions would be accommodated by this structure allowing TCNQ molecules to either be face-to-face with the organometallic rings or by stacking alongside the cation, thereby minimizing void space in the structure.

Sometimes a small change produces very unexpected results.[12] On the left in figure 4.10 is a structure obtained with a TCNQ anion and the bis-arene-iron dication. It has the mixed stack motif of a dication and two anions in a one-dimensional infinite chain. We expected that halogenated derivatives of TCNQ would produce a similar structure, but with either dichloro- or tetrafluoro-TCNQ the structure shown on the right side of figure 4.10 is obtained. This structure is a one-dimensional mixed stack, but the anions are flipped up in a different direction. There was no precedent for such a change previous to these structures. This suggests that interactions between stacks are important, in addition to interactions in the one-dimensional stack. In this case, the halogenation of the central ring pulls more electron density into the ring. This increases the electrostatic interactions between the ring and the highly positive cations in the neighboring stacks. Computer graphic representations of either van der Waals radii or covalent radii show that these structures

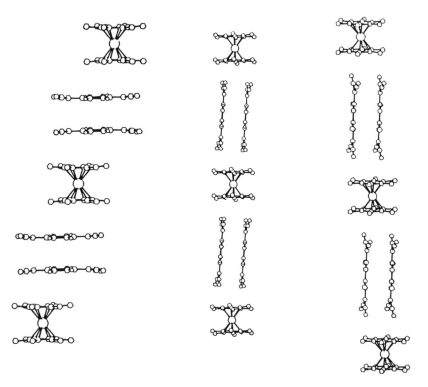

**Figure 4.10**
Stacking motifs for $[(\eta^6\text{-}C_6Me_6)_2M][TCNQ]$ complexes.

are closely packed. We are beginning to understand the charge transfer interactions between the nitrogen atoms of the anions and the rings of the organometallic cation.

This example shows that the structure of a molecular solid can be radically altered by a change in the chemical substituent of one component (the anion). The structure can also be altered by changing the charge on one of the constituents in a compound. To study this situation, we used a different pair of organometallic compounds for cation and anion (figure 4.11). For the cation, we used mesitylene, with methyl groups substituted at every other position of the aromatic ring rather than at every position, as with the cation in the previous example. This compound was aesthetically attractive because of its threefold symmetry. The anion we used, $C_6(CN)_6$, is also a flat, threefold symmetric molecule. Superimposing the two molecules so that the rings overlap gives a very

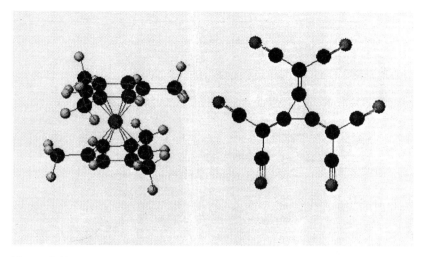

**Figure 4.11**
Structures of metal complex cation $[M(mesitylene)_2^{2+}]$ and $C_6(CN)_6^{n-}$ anion.

nice fit. Calculations based on the orbitals of the two molecules suggested significant electronic interaction between them.

The anion, $C_6(CN)_6$, is available as either a singly or doubly charged anion, allowing us to demonstrate how changing the charge of one of the constituents of a compound can change the structure of the compound. A compound made from a dication and a monoanion gives mixed stacks of alternating dication and monoanion dimer in which the molecular planes of the anion and the arene ligands of the metal cation are perpendicular to each other rather than face to face.[13] The dianion, on the other hand, stacks in a different fashion. It produces an alternating stack compound where the anions approach the ring of the dications, resulting in intense optical transitions due to the charge transfer interactions between the components. The three-dimensional structure of this compound shows the existence of linear chains with an anion next to every cation (figure 4.12). Looking at the structure from the top down (figure 4.13), the anion has six neighboring cation rings and there are some significant nitrogen-hydrogen interactions that appear to be hydrogen bonding in nature. This was the first example that we know of this type of interaction directing the formation of a molecular solid. Six

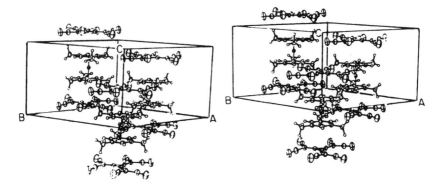

**Figure 4.12**
Unit cell structure determined by X-ray crystallography of the mixed stack
with alternating cations and dianions.

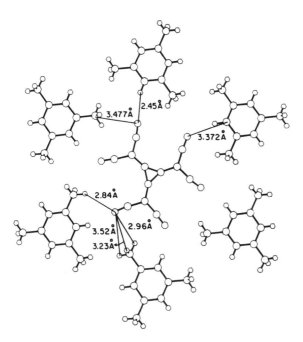

**Figure 4.13**
Top view of the structure in figure 4.12 showing intermolecular contacts in the
two-dimensional sheets of anions and cations.

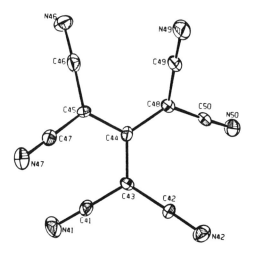

**Figure 4.14**
A nonplanar dianion $[iso\text{-}C_4(CN)_6]^{2-}$.

neighbors around one component is a repetitive theme in the crystallography of organic as well as inorganic compounds.

Figure 4.14 shows a dianion that has $C_{3v}$ symmetry, not $D_{3h}$ symmetry, which is not planar because the arms twist out of the plane. Working with this molecule, we could form donor-acceptor charge transfer complexes. This is contrary to the notion that you need planar constituents to create donor-acceptor complexes.[14] Instead of a linear chain, we actually observe zig-zag chains. As a result of substitution with different metals, and a knowledge of the ionization potentials of the donors and the electron affinities of the acceptors, we can predictably fashion the optical properties of the complex to get a variety of colors, ranging across the visible spectrum.

**Controlling Dimensionality in the Solid State by Electrostatic Templates**

We have seen some examples of point charge molecules that have a single positive charge or a double positive charge localized at one point in space and examples of the types of structures you can create from this kind of material. If we could link these cations together to make a one-dimensional axis, a trigonal plane, or a tetrahedral (that is, a $C_2$)

molecule, what would be the effect on the solid-state architecture? We therefore made a series of polycations that were prepared by methods developed by Paul Fagan using the reaction: $Cp^*RuL3$ + arene → $Cp^*Ru(arene)$ + 3L.

$Cp^*$ describes a fivefold ring called cyclopentadiene ($\eta$-$C_5Me_5$) with methyl groups ($CH_3$ or alternatively designated Me) at each carbon position. Ru is ruthenium. L refers to a ligand of some sort that is labile and tends to be quickly extruded from the compound. Reaction of this precursor with an arene complex (almost any aromatic ring that you can imagine) forms a cation $Cp^*Ru^+(arene)$. This method was exploited to make polycations with different shapes and spatial orientations of charge.[15]

Four examples are shown in figure 4.15. In the upper left is an example of a point charge ($D^+$). The upper ring is the fivefold symmetric cyclopentadiene ring; the bottom ring is an arene ring. In the upper right is a one-dimensional axis of dipositive charge ($D^+$-$D^+$) in which two point charges are hooked together by a cyclophane ring. The distance between the two benzene rings is about 2.85 Å. This molecule is robust and linear. In the lower left triptycene has been modified to get a threefold arrangement of charge. The molecule in the lower right gives us two orthogonal axes of dipositive charge [$(D^+)_4E$]. Essentially it is the structure in the upper right extrapolated to four charges.

Can the electrostatic interactions induced by these unique spatial arrangements of positive charges cause the counterions (the anions) to orient in a specific motif? With the point charge, the crystal packing could be difficult to control and unpredictable. As shown in figure 4.16, we might, by virtue of the propensity of the rings in the organometallics to lie face-to-face, get one-dimensional segregated stacks, in which the polycyanoanions aggregate along an axis parallel to that of the cations. Alternatively, we might get mixed stacks in which the cations and anions alternate along the same axis. Our hope was that, in contrast with this unpredictability, the one-dimensional dipositive charged rod would constrain the planar polycyanoanion acceptor molecules to give one-dimensional segregated stacks, as in figure 4.16. Extending this hope further, the two-dimensional array of positive charge might force a packing as depicted on the right in figure 4.16.

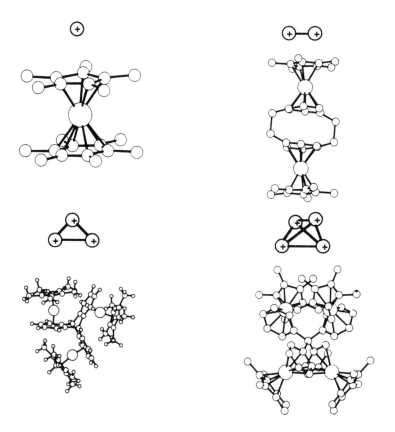

**Figure 4.15**
Organometallic cations of four diverse shapes. The point charge $(D^+)$ is
$[Cp^*Ru(\eta\text{-}C_6Me_6)]^+$. Hydrogen atoms have been omitted from the drawing.
The rod charge $(D^+\text{-}D^+)$ is $[(Cp^*Ru)_2(\eta^6,\eta^6\text{-}[2_2]\text{-}1,4\text{-cyclophane})]^{2+}$. The tri-
angular cation is $[(Cp^*Ru)_3(\eta^6,\eta^6,\eta^6\text{-triptycene})]^{3+}$. The tetrahedral cation
$[(D^+)_4E]$ is $[Cp^*Ru(\eta\text{-}C_6H_5)]_4E^{4+}$, where $E = C$ or $Si$.

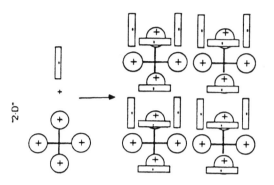

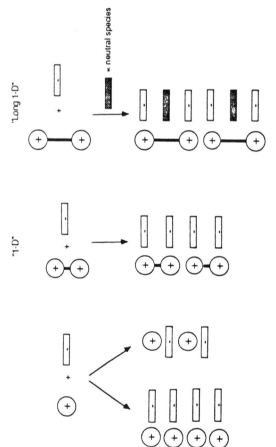

= neutral species

**Figure 4.16**
Schematic representation of different stacking motifs for the ruthenium polycations.

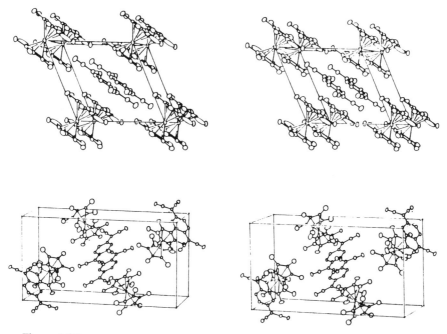

**Figure 4.17**
Crystal structures of the 1-D phase (top) and the dimer phase (bottom) obtained by reaction of the $D^+$ cation with TCNQ anion.

We first looked at the point charge molecules.[16] With reduced TCNQ, either directly mixed with the cation or formed by reduction of TCNQ at an electrode in the presence of the cation, we could form two compounds, designated the 1-D phase and the dimer phase, in which we had stacks of cations and anions.[17] The top of figure 4.17 shows part of a linear chain of alternating cations and anions of the 1-D phase in the mixed stack motif. On the bottom is a structure that is not an infinite chain but instead has a cation, a dimer of anions, and then another cation. In no case, however, using many different approaches, did we see a segregated stack structure. It does not appear to be thermodynamically stable. Contrary to our experience with ruthenium, an analogous iron complex is known to form a segregated stack.

We then looked at the one-dimensional rod $(D^+\text{-}D^+)$ with the two positive charges tethered together by the cyclophane ring. The distance between the rings is 9.95 Å. If TCNQ is reduced at an electrode at a potential of $+0.4$ V, in the presence of this cation, a black solid forms at the electrode. The crystals are about 0.5 cm long and are electronically

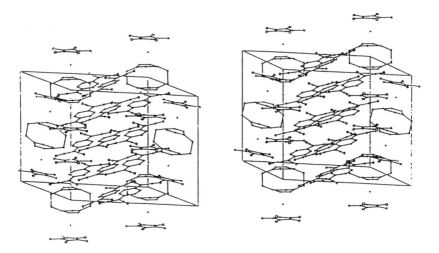

**Figure 4.18**
Structure of the unit cell of $[D^+\text{-}D^+][(TCNQ)2^{2-}]$.

conducting. Crystallographic analysis shows the presence of chains of the dication stacked on top of each other, and chains of anions alongside (figure 4.18). This suggests that the dications, which are stacking together to minimize void space, are enforcing structure on the anions, thus forming a segregated stack structure. The TCNQ anions are present in the crystal in two types of stacks, both have stacking axes parallel to the molecular axes of the cations (figure 4.19). Because of the geometric constraints of the dications, the TCNQ molecules are present in the stacks as tetramers, but because they are aligned in the stack, we get electronic conduction in this material.[18]

The reason why the point charge (mononuclear cation) gives segregated stacks in the case of iron, but not in the case of ruthenium, is related to the distance between the two rings. The distance between the ligands in the ruthenium complex is larger than in the iron complex by about 0.3 Å. This would force the TCNQ anions to move farther apart to stay in registry with the cations in a segregated stack, prohibiting them from overlapping their $\pi$ electron systems to form a one-dimensional chain. However, by tethering the charges together in the dication via the cyclophane rings, and thus squeezing the rutheniums together, a lattice pressure results, allowing four TCNQ molecules to line up along-

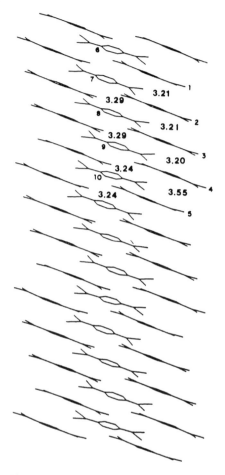

**Figure 4.19**
Representation of the TCNQ stacks in $[D^+\text{-}D^+][(TCNQ)_4{}^{2-}]$ as viewed normal to the 1-D axis. The cations have been omitted for clarity.

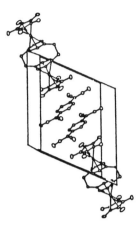

**Figure 4.20**
Unit cell of crystal structure of $[D^+\text{-}D^+][TCNQ^-]_2$ mixed stack structure formed by reduction of TCNQ at $-0.1$ V in the presence of cation.

side one cation. In this case each TCNQ has a nonintegral amount of charge: each has a formal charge of one-half minus. But by designing a system with a known length, which we knew could accommodate TCNQ molecules with a known spacing of 3.3 Å, we could enforce structure and get a one-dimensional segregated stack with desirable electronic properties.

However, the system is not simple. If we change conditions of crystal growth and use a more negative potential ($-0.1$ V) at the electrode, we get a different phase, which has no interesting properties whatsoever. It is a one-dimensional stack with alternating dications and dimer anions in an infinite chain mixed motif (figure 4.20). This illustrates that thermodynamics is not the only factor in crystal growth; one has to control the initial conditions during crystal growth to control the incipient crystal nuclei.

How does electrocrystallization control molecular aggregates and the resulting crystals? We are reducing TCNQ at an electrode to make the anion. From the Nernst equation, $E=E^\circ$, the concentrations of the neutral and anionic forms of TCNQ at the electrode will be equivalent. We propose that under these conditions we can get aggregates (prior to nucleation of crystal growth) of "semireduced" TCNQ molecules, which then crystallize with this dication to form a segregated stack compound with semireduced TCNQ molecules (a tetramer of TCNQ in which each TCNQ moiety has a formal charge of $-0.5$). If, on the other hand, E is very negative, the majority species at the electrode is the fully reduced anion, which we know from spectroscopic investigations forms the dimer anion, resulting in the uninteresting mixed stack phase.

How does a crystal form, and how do we exert control over crystallization? We have demonstrated how to do this thermodynamically. We have two phases that we know are thermodynamically possible. On the other hand, the initial conditions of growth seem very important. For formation of the uninteresting phase (the charge ($\rho$) = 1 phase in figure 4.21), anions form dimers that then form molecular aggregates with the cations to stabilize themselves to a lower free-energy condition. Presumably this is accomplished by minimizing several energetic factors such as the solvation energy. We think that electrostatic charge is also very important because it is a strong force and it exists over long distances. This molecular aggregate eventually evolves into a larger aggregate that eventually forms what we call a prenucleation aggregate, which then proceeds to form the crystalline phase.

On the other hand, with the semireduced phase (the charge ($\rho$) = 0.5 phase in figure 4.21), neutral and anionic TCNQ molecules assemble into aggregates, which line up alongside the dication to form a molecular aggregate. Energy is minimized by the interaction of opposite charges, and these chains evolve into the crystal structure via the $\pi$-$\pi$ interactions between TCNQ molecules.

Can we use this strategy for higher dimensions? We would like to be able to use the two-dimensional tetracations and see the same effect. Can we get an orthogonal packing of the anions? Experiments were done using $C_6(CN)_6$ (figure 4.12) as the anion instead of TCNQ. We first demonstrated that, with $C_6(CN)_6$ as the anion (as with TCNQ as the anion), the dication enforces a linear chain, segregated stack structure on the anions (not shown). If we try in this system the tetracation $(D^+)_4E$ (figure 4.15), which has a two-dimensional distribution of positive charge, we see that the anions are forced to align in two mutually orthogonal stacks parallel to the orthogonal axes of the tetracations (figure 4.22). Thus it appears that these tetracations also enforce structure and that this approach is a fairly reliable way to design molecules.

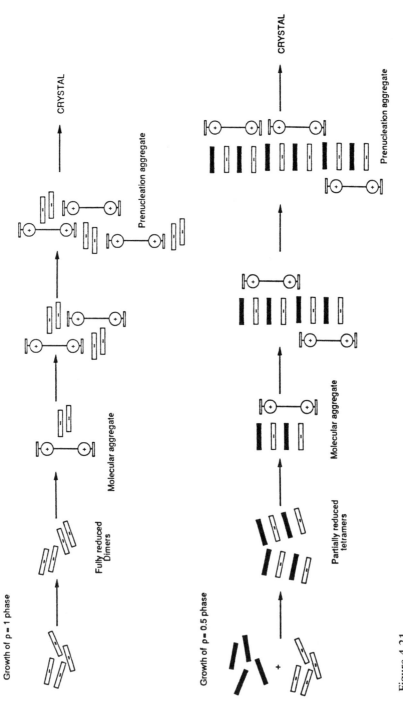

Figure 4.21
Proposed kinetic mechanisms for structure enforced crystallization.

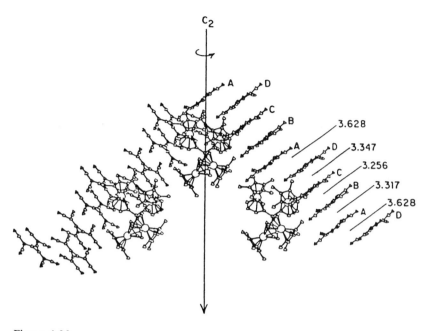

**Figure 4.22**
Representation of two orthogonal $C_6(CN)_6^{1-}$ anion stacks with their nearest $(D^+)_4E$ tetracation.

The two orthogonal stacks of the anions clearly reflect the initial symmetry conditions imposed by the tetracation.

We have extended these concepts to other systems besides these planar polycyanoanions. Hydrogen bonding, as noted, is becoming an important factor in the construction of solids. The dication can enforce the structure of hydrogen-bonded chains that can interact by virtue of hydrogen bonds between the carboxylate groups of a (para) aromatic dicarboxylic acid (figure 4.23). The structure formed has chains of the aromatic dicarboxylic acid going in one direction, which are piled up to form sheets in the perpendicular direction. The sheets are tied together by carbon-hydrogen-oxygen interactions that appear to be hydrogen bonding in nature. These chains have a puckered motif, apparently to minimize void space by pinching around the dication. In a way, this is a very predictable structure. Our rational approach appears to be working for this class of solids as well.

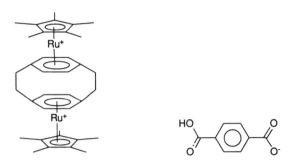

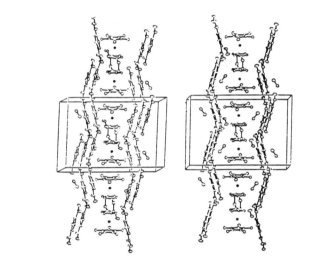

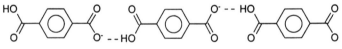

**Figure 4.23**
Electrostatic enforcement in hydrogen-bonded solids.

## Where Are We Going?

We would like to use a "tinker toy" set of some kind to make anions as well as cations of all different spatial distributions of charge to see if we can form solids that have very open structures that might act as hosts for guest molecules. We would like to further develop the idea of electrostatic enforcement along the lines illustrated in figure 4.24. Cations with particular geometric shapes and distributions of localized charges (top panel) could be condensed with appropriate anions. For example, the condensation of the tetrahedral tetracation with a rod-like dianion could be expected to lead to the three-dimensional adamantane-like framework shown in the bottom panel.

One thing we are looking at that has some relevance to nanotechnology is the ability to make functional materials (as I have stated, some of these low-dimensional solids are conducting), but do it at a very small scale—maybe not nanoscale but certainly microscale. For example, there is an electronic conductor called tetrathiofulvalene bromide that can be grown on an electrode. We can grow one-micron-thick conductors. We would also like to direct this growth in some predictable fashion to take advantage of the electronic characteristics and to possibly form patterns of crystals. These are extremely good metallic conductors, and I think that there is a lot of hope for making microscale structures with these materials.

In conclusion, we are inferring the structures of initial molecular aggregates from the structures of crystal lattice that result and are now trying to build up crystal lattices by controlling the structures of the aggregates. If we can ever stop the crystallization process at the nucleation aggregate stage, then we may truly have a way to make nanoscale structures with functional properties. The real key is going to be, what is the relationship between a crystal lattice and these aggregates, and do the electronic properties of the crystalline material necessarily reflect that of the aggregates. We are currently working on ways to stop those processes using these very small-scale structures. We hope to be able to report on that in the future.

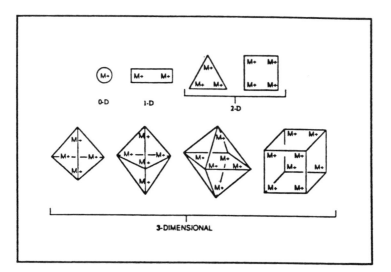

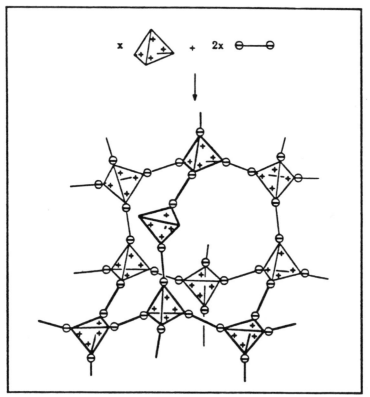

**Figure 4.24**
Construction of poly-ionic lattices employing a "tinker toy" set.

# Notes

1. See M. Ward, "Electrochemical aspects of low-dimensional molecular solids," *Electroanalytical Chemistry: A Series of Advances,* vol. 16 ed. A. Bard (New York: Marcel Dekker, Inc., 1989). This paper discusses some of the electronic properties of these materials.

2. G. Desiraju, *Crystal Engineering: the Design of Organic Solids,* Materials Science Monographs, vol. 54 (Amsterdam: Elsevier, 1989). In this text, crystal engineering is defined as "the understanding of intermolecular interactions in the context of crystal packing and in the utilization of such understanding in the design of new solids with desired physical and chemical properties."

3. The reactions studied by Schmidt "were termed topochemical since they were considered to occur with a minimum of atomic or molecular movement" (Desiraju). Whether or not a reaction occurred depended upon the geometry of the molecules in the reactant crystal, rather than upon molecular diffusion. Consequently, the structure of the reaction product would closely resemble the crystal symmetry of the reactant. Typically this resemblance would result in fewer products being formed by a given topochemical reaction than by a corresponding reaction in solution.

4. See Desiraju, *Crystal Engineering.*

5. See Ward, "Electrochemical aspects," 184–188, for a more complete discussion of charge transfer in low-dimensional molecular solids, from which the following has been condensed.

The crystallization of (planar) aromatic species can lead to the face-to-face stacking of molecular species such that the interactions of their $\pi$ electron systems can result in electron delocalization and subsequent stabilization of the solid through the energy gained by delocalization. This alignment of the molecules in the solid commonly is associated with charge transfer interactions in which charge is transferred from the HOMO (highest occupied molecular orbital) of the electron donor species (D) to the LUMO (lowest unoccupied molecular orbital) of the electron acceptor species (A). Thus most charge transfer solids possess donors and acceptors that are at least partially charged ($D^+A^-$). The net amount of charge transferred is not necessarily integral, and in the case of the segregated stack material TTF-TCNQ, the average charge is $TTF^{0.59+}$ $TCNQ^{0.59-}$.

6. O. Ermer, "Fivefold-diamond structure of adamantane-1,3,5,7-tetracarboxylic acid," *J. Am. Chem. Soc.* 110 (1988): 3747.

7. See Desiraju, *Crystal Engineering.*

8. See Desiraju, *Crystal Engineering,* 186–192. The role of halogen atoms is described in producing "beta structures" characterized by the stacking of planar, dichlorosubstituted aromatic molecules to give close packed, highly overlapped and parallel structures with a separation between planes of about 4 Å. To adopt such structures, molecules typically must lack other, stronger intermolecular interactions.

9. See Desiraju, *Crystal Engineering,* 142–165.

10. The Madelung constant is a parameter used in the calculation of the Coulombic energy of a crystal lattice that results from the balance of attractive interactions of opposite charges—and the repulsive interactions of like charges—in the lattice. The Madelung constant reflects the geometry of the arrangement

of the charges in the lattice and is thus specific for lattice type and independent of lattice dimensions.

11. This structure is a dication because the metal ion complexed in the center has a charge of 2+. This charge is spatially a point charge (zero dimensional).

12. M. Ward, D. Johnson, "Electrocrystallization and structural and physical properties of charge-transfer complexes derived from $[(\eta^6\text{-}C_6Me_6)_2M]^{2+}$ (M = Fe, Ru) and TCNQ (TCNQ = tetracyanoquinodimethane)," *Inorganic Chemistry* 26 (1987): 4213–4227.

13. M. Ward, "Linear chain organometallic donor-acceptor complexes and one-dimensional alloys. Synthesis and structure of $[(\eta^6\text{-}C_6Me_3H_3)_2M]$ $[C_6(CN)_6](M = Fe, Ru)$," *Organometallics* 6 (1987): 754–762.

14. M. Ward, J. Calabrese, "Organometallic donor-acceptor complexes with non-planar donors: the zigzag linear chain complex $[(C_6Me_6)_2M^{2+}][iso\text{-}C_4(CN)_6^{2-}](M = Fe, Ru)$," *Organometallics* 8 (1989): 593–602.

15. P. Fagan, M. Ward, J. Calabrese, "Molecular engineering of solid-state materials: organometallic building blocks," *J. Am. Chem. Soc.* 111 (1989): 1698–1719.

16. M. Ward, P. Fagan, J. Calabrese, D. Johnson, "Electrostatic structural enforcement in low-dimensional solids: synthesis, structure, and electronic properties of polycationic ruthenium complexes with polycyanoanions," *J. Am. Chem. Soc.* 111 (1989): 1719–1732.

17. Mixing solutions of cation and reduced anion gave a green crystalline solid of the 1-D phase as the majority product and a purple crystalline solid of the dimer phase as the minority product. Reduction of the anion in the presence of cation gave only the 1-D phase. (ibid., 1721.)

18. The fact that there are two crystallographically different types of TCNQ stacks in $[D^+\text{-}D^+][(TCNQ)_4^{2-}]$ leads to electronic localization that complicates analysis of the excited states of this complex and its conductive and magnetic properties. See J. Morton, K. Preston, M. Ward, P. Fagan, "Single crystal electron paramagnetic resonance spectra of triplet excitons in $[(C^*pRu)_2(\eta^6,\eta^6\text{-}[2_2]$ (1,4)cyclophane)$^{2+}][(TCNQ)_4^{2-}]$," *J. Chem. Phys.* 90 (1989): 2148–2153.

19. See K. Harris, M. Hollingsworth, "Organic crystals: Losing symmetry by design," *Nature* 341 (1989): 19.

Noncentrosymmetric organic solids have very desirable practical properties, such as nonlinear optical responses. This article discusses advances in the "crystal engineering" of hydrogen-bonded structures that were reported by M. C. Etter of the University of Minnesota at the 9th International Conference on the Chemistry of the Organic Solid State, Como, Italy, 2–7 July 1989. Etter et al. have studied the crystallographic data in the Cambridge Structural Database and developed a set of empirical rules for the hydrogen-bonded linkages formed by different functional groups. These rules enable predicting the crystal packing of molecules with appropriately arranged donor and acceptor groups for hydrogen bonds.

The work of Etter's group has focused on frequency-doubling (second-harmonic generation), by which a crystal converts light of one frequency to light with twice that frequency. This ability depends not only on various electronic features (particularly a parameter termed the second-order hyperpolarizability tensor) of the individual molecules, but on their spatial arrangement in the crystal, particularly upon the absence of a center of symmetry. Unfortunately, many molecules with excellent electronic properties crystallize spontaneously in centrosymmetric structures. Hence the approach is to arrange hydrogen bond donors and acceptors in positions that promote noncentrosymmetric aggregation. Partial success has been reported with co-crystals of *p*-aminobenzoic acid and 3,5-dinitrobenzoic acid.

## Discussion

*Drexler:* As you know, I think it is a very interesting approach to nanoscale structures to take relatively complex molecules, of the sort that you have been working with, and to make them side groups on a polymer chain so that you could have both the electrostatic constraints that you are speaking of here and the constraints associated with a covalent backbone. Would you care to comment on the potential for that combination?

*Ward:* That is certainly possible. We have done a little bit in that direction, but not a lot. If you will recall the structures that I showed with the bis-arene and dianion structures in mixed stacks, we've made polymers with molecules appended to them which are commonly known as paraquat. This also forms a mixed stack complex with this dianion. We can form materials which are polymeric in nature but actually have donor and acceptor stacked as alternating pairs alongside the backbone. I think therefore that it is possible, but limited by what the organic chemist can put together. Most of our focus has been on crystallographic analysis and trying to understand the relationship to structure. This is not very easy to do with the polymer system, although it is certainly a possibility because of all the "hooks" and "tethers" and "knobs" furnished by the large number of functional groups that the organic chemist has at his disposal.

*Fahy:* You spoke about the desirability of being able to control crystallization, to limit it at one point, or to direct it all in one direction in another. I am a low-temperature biologist. In nature there are examples of inhibiting ice crystal growth with natural proteins that absorb into the surface, so there might be some possibilities for designing molecules that you might add to your brew at the right moment to keep the nuclei that you form from growing. The other point is that in the field of metallurgy there are techniques for directional solidification involving the imposition of thermal gradients—design a stage and advance your crystals through that. So I think there are good prospects for solving some of the problems that you are talking about.

*Ward:*   The first case you mentioned seems to be a very clear analogy. The second seems to me to be more a macroscopic engineering approach. That may be equivalent to our electrochemical control of structure here, controlling the initial conditions and such, but it is not a "bottom up" approach.

*Audience:*   I'd like to comment about something that you said in the beginning. You implied that crystallization was the same thing as self-assembly. In biology, the term self-assembly implies more the assembly of different structures that come together to form closed structures. I think that these are two elements that are not usually included in the term *crystallization* as used by chemists.

*Ward:*   You're right. Chemists just hadn't thought of self-assembly. They are now because they are starting to consider very complex systems, ternary systems that are different in shape and size. Right now there are two different classes of work. Crystallographers have spent a lot of time looking at single-component systems, which don't, by your comments, relate to self-assembly. But there is a smaller class of two-component systems that are much more complex and are just now being investigated.

*Kantrowitz:*   Is it possible to influence the formation of crystal with external fields, for example, intense laser fields?

*Ward:*   That's a good question. We've thought about that. You can grow magnetic phases in the presence of strong fields. We have tried to grow crystals in strong electric fields, but only in the sense of trying to control the direction of growth. We haven't really looked at phase selectivity as a function of applied magnetic fields. But it is a really good point. Some of these phases tend to have spin states. For example, the one-dimensional complex that I showed actually has triplet spin excitons.[19] Perhaps we should try to confer some selectivity based upon magnetic properties.

*Kantrowitz:*   The easiest way to get a very powerful field is with a laser.

*Audience:*   Can you comment on noncentrosymmetric crystal structures?

*Ward:*   That is at the heart of crystal engineering. To get high $\chi^2$ properties—the second-order tensor responsible for frequency doubling in optical materials—the approach has been to start with molecules that have functional groups that will result in a noncentrosymmetric structure. I showed some of the nitroanilines, for example, which have a head-to-tail hydrogen bonding pattern so that they will line up in a noncentrosymmetric fashion. For $\chi^3$ effects, it's not so obvious what properties we need in those crystals. Many groups have very intensive efforts in trying to develop $\chi^2$ and $\chi^3$ material from crystalline phases.

*Drexler:* In systems engineering, one is very concerned with the tolerance of components for a variety of conditions that will be encountered during (to use the macroscopic engineering terms) fabrication and incorporation into the operational environment. Could you comment on the stability of these building blocks and the aggregates of them under various conditions of chemical environment, solution environment, thermal stability of crystals, and so on?

*Ward:* Our building blocks were chosen because they are very robust in order to facilitate synthesis and the measurement of their properties. The compounds that I've shown you are extremely air-stable; they are thermally robust up to reasonable temperatures. Part of the problem with organics and organometallics is that high temperatures (100° C) will always be a problem. With crystals, you always have to worry about dissolving the crystal that you've just grown. The compounds that we have chosen have been highly charged, and they have the solubility properties of a brick—they will not dissolve once they have formed. They are insoluble in the medium in which they are grown so they crystallize spontaneously at electrodes. The compounds that I have described today are indicative of desirable properties thermally and in solubility. I haven't shown you the ones that don't have desirable properties in that context. The organic chemist has a lot of tools to design molecules with good properties. For example, you can consider making systems based on hydrophobic sheets that will not be touched by aqueous media, or vice versa.

*Audience:* You mentioned that one experiment involved paraquat. How do you protect yourself against inadvertently generating something that is hostile to you?

*Ward:* We're very careful about our working conditions and how we handle toxic materials. Paraquat has been ubiquitous in electrochemistry laboratories and hasn't harmed anyone yet.

# 5

# Molecular Engineering in Japan: Progress toward Nanotechnology

Hiroyuki Sasabe

When we think of molecular electronics, we think of three approaches. The first is materials, both synthetic and natural, the latter coming mostly from biological materials. The second approach is techniques to organize these materials, these molecules. Many different techniques can introduce this organization, but mostly we study the Langmuir-Blodgett technique and liquid phase epitaxy, or, for a dry system, molecular beam epitaxy or ion beam sputtering. Third, to make real molecular devices, we use lithography techniques. If in the future we can make structures by self-assembling molecules, then we will not need lithography, but until then we will need lithography techniques, such as electron beam or ion beam lithography.

There are two approaches to realize molecular electronics. One is to make ever smaller devices—the top-down approach. Our approach is, however, from the bottom-up—building larger structures from molecules to realize functional devices.

In the case of biological materials, this approach means the use of proteins and enzymes and requires protein engineering. This can involve three-dimensional crystals of proteins, but even two-dimensional arrays of protein molecules can be important. The full variety of living materials can also be directly exploited. Neural networks provide a good example of the usefulness of biological models for data processing systems. Synaptic simulations are also an important model. Finally, lithographic techniques for device processing can include STM (scanning tunneling microscopy) and AFM (atomic force microscopy).

In Japan, there are many national projects for molecular electronics. As you may know, bureaucratic sectionalism is quite strong in Japan.

Three major ministries are involved: the Science and Technology Agency (STA), the notorious Ministry of International Trade and Industry (MITI), and the Ministry of Education, which supports fundamental sciences in universities. STA runs many programs; I will mostly introduce the Frontier Research Program, which is prominent in our institute. This program began in 1986 and is intended to run for 15 years. Other efforts supported by ERATO (Exploratory Research in Advanced Technology) include programs in functional polymers, ultrasmall structures (e.g., motors), and supermolecular structures.

An international creative science and technology program began in 1989 including a program on artificial organs and biocompatibility. This is a joint program between the Tokyo Medical Institute and the University of Utah.[1] An "intelligent materials" program began in 1990 organized by Professor Miyata of the Tokyo University of Agriculture and Technology. Intelligent materials refers to an imitation of biological systems to produce various functions as the result of inherent properties of the materials themselves, such as self-assembly, self-healing, and molecular recognition.

MITI coordinates several large projects, especially International Science and Technology for the Next Generation. In addition to molecular electronics, they have a program on super lattices for three-dimensional devices. The main focus is on semiconductors. Projects to study nonlinear optics and optical devices using organic molecules have just recently begun. There is also a bioelectronics device project that started in 1987.

In the Frontier Research Program, we have ten projects now in three groups. One focuses on biohomeostasis, one is on frontier materials, and the most recent program studies the brain and neurosciences. There are three research teams in the materials program. One is working on quantum materials, especially for quantum devices and one- and two-dimensional quantum states. The second team is working on nonlinear optics and advanced materials, but in Japan this is called the laboratory for molecular devices. The third team is the laboratory for bioelectronic materials and bioelectronic devices. This project is quite international.

This chapter presents recent work with proteins done at the laboratory for bioelectronic materials. The main effort of this group is to create two-dimensional crystals of protein molecules. As a model protein, we

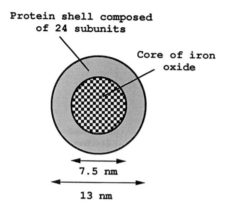

Protein shell composed
of 24 subunits

Core of iron
oxide

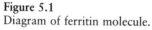

7.5 nm

13 nm

**Figure 5.1**
Diagram of ferritin molecule.

selected ferritin, which is a quite spherical protein as shown in figure 5.1.[2] We applied several techniques, but the simplest was the Langmuir-Blodgett technique, in which an aqueous solution of the protein is spread over the air-water interface and then transferred onto a solid substrate. Figure 5.2 shows a scanning electron microscope (SEM) picture of the ferritin molecules transferred onto silicon wafers. It is apparent that the packing is not solid. Even if we change the surface tension from 10 to 50 dynes per centimeter, the packing is not very tight.

To achieve better packing, we tried an adsorption technique. The ferritin molecules were dissolved into a surface covered with a polypeptide monolayer. The ferritin molecules were adsorbed to the polypeptide monolayer because of electrostatic interactions. Figure 5.3 shows that quite good packing of the molecules is obtained, but there are still numerous defects and dislocations.

Dr. Hara in our group has observed (using the STM) very nice packing of photoactive enzyme, nitrile hydratase, prepared by means of adsorption onto a graphite substrate (plates 9, 10).

Another protein that we have studied is cytochrome b562, which has a heme group supported by four α helices (figure 5.4). Using computer modeling we determined the distribution of electrons and the resulting positively and negatively charged regions in the protein. We are using protein engineering techniques to change the electron distribution to try

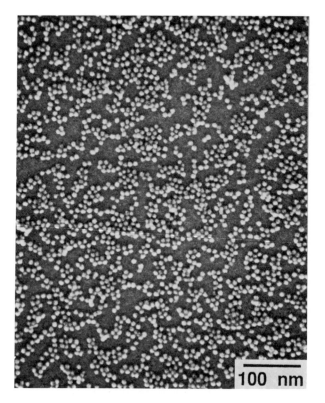

100 nm

**Figure 5.2**
Langmuir-Blodgett film of ferritin molecules on silicon wafer.

to improve the packing of the molecule as analyzed by adsorption techniques.

We are also studying cytochrome c3. We first used a computer graphic system to study the possible rotational orientations to determine which might be most favorable for adsorption to a substrate (plate 11). We then used the STM to look at a single molecule on a substrate (plate 12) and found the STM pattern to be very similar to our expectations. To our surprise, the size as well as the pattern was exactly as expected. Techniques for aligning molecules on surfaces are quite important, but observational techniques to characterize these molecules are also very important.

The next protein that we studied was bacteriorhodopsin. (See Dr. Birge's chapter, this volume, for his explanation of the workings of this

**Figure 5.3**
Adsorption of ferritin molecules onto a polypeptide monolayer.

protein.) Bacteriorhodopsin contains seven helices. Our STM observations of bacteriorhodopsin (plate 13) reveal a uniform distribution of the protein molecules in the patch of purple membrane and helical structures of bacteriorhodopsin at the edges of the membrane.

There are stable states in the photochemical cycle of bacteriorhodopsin, but there is also an intermediate state with a very short half-life. We have tried to prepare a mutant bacteriorhodopsin with a more stable intermediate state. Usually the retinal chromophore is bound to amino acid residue 216 through Schiff base and extended to residue 96 (aspar-

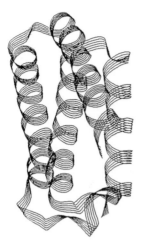

**Figure 5.4**
Schematic structure of cytochrome b562.

tate). In a mutant, aspartate 96 has been changed to asparagine. This change caused a thousandfold increase in the half-life of the intermediate state, from several milliseconds to several seconds, potentially a quite advantageous change for use in optical memories.

We have also studied how to align molecules in a dry state using molecular beam epitaxy techniques that have been especially designed for organic molecules by modifications to permit work at liquid nitrogen temperatures and for vapor pressure control of organic molecules. We used molybdenum disulfide as a substrate, and the organic molecules were derivatives of phthalocyanine. Using this apparatus, we can make flat crystals of phthalocyanine (figure 5.5). These are easily recognized by the SEM patterns. A comparison of surfaces made by organic molecular beam epitaxy (MBE) and by conventional vapor deposition is shown in figure 5.6. No defects can be observed in the very smooth surface produced by organic MBE.

While the surface is growing, it can be observed by reflection high energy diffraction (RHEED) patterns. The fundamental 2.7 Å streaks of the substrate can be observed, and another streak of 12 Å can be observed during the deposition of the phthalocyanine (figure 5.7).

As an example of nonlinear optics, we designed the molecule dicyano vinyl anisole. We can easily crystallize this molecule and check the second

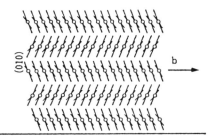

**Figure 5.5**
Crystals of phthalocyanine on substrate made by MBE.

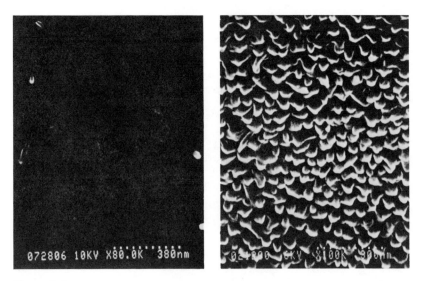

**Figure 5.6**
Surfaces of phthalocyanine made by organic molecular beam epitaxy (left) and by conventional vapor deposition (right).

**Figure 5.7**
Changes of RHEED patterns due to deposition of phthalocyanine.

harmonic generation. This produces green light by Nd-YAG laser excitation (1064 nm). A much shorter wavelength exciting light (812 nm) produces blue-violet light. In this way we can design molecules for nonlinear optical materials. Plate 14 shows an STM of dicyano vinyl anisole molecules. The exact pattern of the molecules can easily be recognized.

In conclusion, to realize molecular electronic or bioelectronic devices requires progress in three areas: (1) materials, (2) controlling molecular alignment, and (3) lithography techniques. These three should be developed in parallel, but the first two are of immediate importance.

## Notes

1. Unfortunately this program was canceled.
2. Ferritin is a large protein with a molecular weight of about 460,000 daltons that stores iron in tissues. Its large internal cavity (about 80Å diameter) can hold some 4,500 ferric ions. See L. Stryer, *Biochemistry,* 3rd ed. (New York: W. H. Freeman and Company, 1988) 299, 595.

## Discussion

*Audience:*   In your work with phthalocyanine surfaces, what was the orientation of the plane of the phthalocyanine molecule with respect to the surface? Was it parallel?

*Sasabe:*   Initially parallel, but the orientation changes as the surface becomes thicker.

*Audience:*   When you evaporate on the molybdenum sulfide, do you evaporate on the 1-1-1 molybdenum or the 1-1-1 sulfur?

*Sasabe:*   The 1-1-1 sulfur is on the surface.

*Schwartz:*   I assume that the rest of you were similarly impressed as I was with what appears to be the beginnings of a rather coherent program of development and research in Japan—and perhaps somewhat envious as well.

# 6

## Strategies for Molecular Systems Engineering

K. Eric Drexler

Most of the chapters in this volume focus on several enabling technologies for molecular manufacturing and molecular nanotechnology. These fall into the Chemistry and Molecule-positioning boxes of the diagram shown in figure 6.1. Bill Joy's discussion of future computing systems, in contrast, describes not an enabling technology but a potential application of advanced molecular technologies. (Improvements in software and hardware for the computer aided-design of molecules, however, are in the former category.)

Chemistry is fundamental to molecular nanotechnology. However, I am not a chemist by training. I am an engineer attempting to understand how various technologies and capabilities can fit together to build useful systems. As Herbert Simon suggests, in his classic essay, "The Architecture of Complexity,"[1] this is best done by treating technologies as black boxes—where possible—peeling away layers of abstraction to understand more of their behavior as necessary. Thus, the chemistry behind the ideas in this chapter should be seen as chemistry as understood by an engineer who has studied the subject from a somewhat peculiar perspective.

This chapter first presents some advances in chemistry and biochemistry that are leading toward molecular systems engineering. Second, I describe prospective developments at the level of molecular protoassemblers (see figure 6.1), combining molecular engineering with capabilities from micropositioning, exploiting the technology of atomic force microscopy. The third segment offers a brief look at the topic of molecular nanotechnology based on molecular assemblers. The technological goal of molecular nanotechnology is the ultimate focus of this volume and

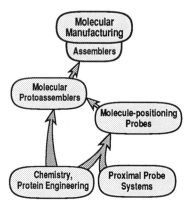

**Figure 6.1**
Developmental pathways in molecular systems engineering and nanotechnology.

provides the motivation for the chapters that consider the possible long-term consequences of this goal.

## Early Molecular Systems Engineering

### Working with Folding Polymers

In molecular systems engineering, there are strong advantages to working with molecules that resemble biological macromolecules, that is, polymeric structures that fold to form specific three-dimensional structures capable of intermolecular self-assembly. Existing machines can synthesize DNA; others can synthesize proteins. Furthermore, proteins that consist exclusively of the 20 genetically encoded amino acids can be made using biotechnology, by synthesizing genes and expressing them in bacteria. My focus here, however, is to describe some of the advantages of engineering proteinlike molecules using a wider range of building blocks than these 20 amino acids.

To begin with, it may be useful to ask why folding polymers are useful. The example of proteins shows that (1) they can contain hundreds to thousands of atoms, (2) they can form solid objects, (3) they can be as stiff as plastic or wood, and (4) they can be used to build machines. It is well known that one can build mechanical devices on a macroscale using plastic or wood; likewise, one can build mechanical devices on a nanometer scale with materials of similar stiffness. And indeed, it is clear that folding polymers can form molecular machines because we see such machines in biology.

Folding polymers have several advantages from an engineering perspective.

1. Modular structure. Engineers like to have standard parts that they can put together in systematic ways to make a wide range of structures.
2. Systematic assembly of modules. Merrifield synthesis assembles a complex polymer chain step by step and is an established technology.[2]
3. Proven adequacy as a basis for constructing molecular machines.

### Self-assembly

The basis of molecular self-assembly is complementary surfaces and fast, spontaneous motion. For molecular parts to self-assemble, they must have surfaces that match in shape and in electrostatics (as emphasized by Michael Ward, chapter 4, this volume); other interactions can also be significant. From an engineer's perspective, a key to self-assembly is that molecules move quickly and don't wear out as a result. Brownian motion in aqueous solution makes protein-sized parts shift by their own diameter roughly $10^6$ times per second, they turn by a radian roughly $10^7$ times per second, and they shift by an atomic diameter $10^{10}$ times per second. Thus they thoroughly and rapidly explore their environment and possible arrangements, becoming trapped when details of surface complementarity conspire to provide a deep energy well. The result can be rapid, reliable self-assembly of complex structures.

### First-order Analysis of Protein Design

Regarding protein design, the state of thinking in the early 1980s was rather pessimistic. It was generally assumed at the time that designing a structure that would fold properly required the ability to predict how a given structure would fold, that is, how natural proteins would fold, given knowledge only of their amino acid sequence and no knowledge of the folded structure of a similar protein. The energetic difference between a folded and an unfolded protein, however, typically amounts to only 10–25 kT at room temperature, a difference too small to calculate from first principles in so large a structure. Further, the torsional flexibility of the protein backbone and the side chains creates a combinatorially huge search space of possibilities. (Even a short chain of only 100 amino acid residues, with five states per residue, generates on the order of $10^{70}$ conformational states.) With marginal stability and so many possible conformations, predicting the one stable fold seemed hopeless, and it was therefore assumed that fold design was also hopeless.

In 1981 I published a paper in the *Proceedings of the National Academy of Sciences* arguing that protein design was a fundamentally different problem than the prediction of the folding of natural proteins, and noting that design was arguably easier because one could design for more predictable folding to more stable structures than those seen in nature.[3] In essence, the argument was that natural selection has no pressures driving it to achieve engineering objectives, and so evolution can, by these standards, be improved upon. Tracy Handel (chapter 3) describes work in Bill DeGrado's lab;[4] we see that successful design has been accomplished and that the products are indeed more stable than those found in nature. I take this as a confirmation that abstract engineering analysis can be successfully applied to the molecular domain.

**Improving Fold Control**
Similar arguments suggest that one could gain better control of the folding of macromolecules (both in stability and predictability) by using a wider range of building blocks than the 20 genetically encoded amino acids. Several researchers have taken steps in this direction. Manfred Mutter, for example, has used branched backbone structures instead of strictly linear structures to achieve better control of folding, and others have used right-handed amino acids and other unnatural monomers.[5] It may be that what I am about to describe will be unnecessary. On the other hand, my objective in much of this work is to set lower bounds on future capabilities, and this is best accomplished by constructing robust arguments and by seeking multiple ways to accomplish goals. Accordingly, I would like to argue that even if success with natural proteins does prove too difficult, we could still succeed in engineering macromolecular objects from folding polymers by building on the technology base of protein chemistry. How can nonstandard amino acids and backbone structures help?

1. Backbone and R-group rigidity. Backbone side chains can be made more rigid. This reduces the number of conformations that a structure can have in solution, thereby entropically favoring the folded state. It also favors a specific folded state, and hence should reduce problems of misfolding and aggregation—the banes of work with modified proteins.

2. Steric and electrostatic diversity. Designs can exploit side chains with a greater diversity of shapes and electrostatic properties. Having more choices can help in designing the combination of closely fitted puzzle pieces that form the core of a protein.

3. Pairwise matching of side groups. This can strongly direct folding toward a particular configuration. Different parts of the chain can be linked before or during folding by a selective interaction, similar to the pairing that we see in DNA (or a stronger pairing mechanism). This can radically constrain the number of conformations that the system can assume, again stabilizing a particular fold, increasing stability, and avoiding numerous undesirable behaviors.

4. Cyclic backbone structures. This involves constructing covalent loops during synthesis, with benefits similar to the above.

5. Staged crosslinking and complexation. I have discussed elsewhere several proposals, including one for using changes in redox conditions to encourage the selective cross-linking of thiols to form disulfide bridges similar to the cystine bridges in natural proteins.[6] Successive sets of bridges (joining thiols of decreasing pKa) would tend to form under successively more oxidizing conditions.

These approaches provide additional strategies for controlling polymer folding and thus constructing stable, nanometer scale, atomically defined objects. Whether by using these strategies or by employing better design methodologies for standard proteins, it appears that we can learn how to make molecular objects. Further, designing polymers to fold correctly is quite similar to designing them to undergo self-assembly. From the standpoint of design, the difference is slight between having a loop of a chain stick to other loops of the same chain and having it stick to loops on a separate chain.

## Molecular Manipulators

### Capabilities of Assemblers

The above argues that we should be able to make precisely structured building blocks of multinanometer dimensions, and that we should be able to get them to self-assemble to form larger systems. Here I will briefly outline some applications of molecular engineering capabilities to the construction of tools that can further extend molecular engineering capabilities. The central concept is that of a molecular assembler, a device capable of performing positionally controlled synthesis.

Currently, chemistry uses reactions that occur spontaneously when molecules encounter each other by diffusion, colliding in all possible positions and orientations. Diffusive encounters also underlie macromolecular self-assembly, which can be highly specific and reliable. However, chemical synthesis typically involves smaller molecules, which are intrinsically less selective. Larger molecules have more surface area and more distinct features; their pairwise interactions can be more selective. Chemists have cleverly exploited small energy differences between different reaction pathways and can achieve enough specificity to assemble molecular objects requiring tens and sometimes hundreds of sequential steps. The results are impressive and seem sufficient to provide the tools necessary to move on to more advanced capabilities. It is worth considering what those capabilities will be so that we can see how to develop them and what they have to offer.

1. Able to precisely position molecules. The first capability is positional control of reactive molecules, bringing them together in specific configurations much as enzymes do. This can provide the kind of specificity familiar in enzyme reactions.

2. Programmable. Rather than an enzyme molecule that can bring together only a few kinds of molecules to cause only a single kind of reaction, we want a device that can perform programmable positioning. Such a device resembles, in some respects, a macroscopic industrial robot.

3. Able to use familiar chemistry. Such a system, by positioning reactive moieties with respect to one another, would be able to guide familiar chemical reactions in a site-specific way, essentially by creating localized regions of high effective concentration.

4. Able to use exotic chemistry. In the longer term, it will be possible to use more exotic chemistry, but this is really not applicable to the discussion of molecular manipulators; it is instead relevant to more advanced developments in assembler chemistry, where reaction environments are subject to more thorough control.

## Elements of Molecular Manipulators

A *molecular manipulator* is a proposed device (or class of devices) that will give positional control of chemical reactions, but without the generality that is described in *Engines of Creation* and the bulk of my writings. From a systems engineering perspective, the essential elements of such a device are the following:

1. Product-structure binding mechanism. If we are working on a product structure and want positional control of reactions on its surface, we need a way to bind it to a mechanically stable support.

2. Reagent or reagent-complex positioning mechanism. This is a second essential element: a means for positioning the reactive moieties. This could involve either a reactive molecule bound directly to the positioning mechanism or one bound to a complex that is in turn bound to the positioning mechanism. If the binding is a reversible, equilibrium process in solution, then the reactive device attached to the positioning mechanism can be changed by changing the composition of the solution. As a solution washes over a mechanism, there is a certain probability (per unit time) that the reagent structure comes off, and a certain probability (per unit time) that a new reagent structure takes its place, binding from the solution. This could be a different structure, capable of different reactivity.

3. Reagent-complex family. To do a wide range of chemistry, we need a family of different reagents, each of which has some feature in common (perhaps after complexation) that enables it to bind to the same support structure.

## Concepts for a Molecular Manipulator

An abstract concept for a molecular manipulator is illustrated in figure 6.2. It includes a mobile structure that can be positioned with the ~ 0.1 Å accuracy now familiar in scanning tunneling microscopy (STM) and atomic force microscopy (AFM) work. At the end of this positioning structure is a standard interfacing structure, such as a molecule permanently attached to a tip. Bound to that molecule is a transient tip molecule, which might be a reagent bound to a complex (to provide broad options for the interface design). In an alternative implementation, the reagent itself would include a standard interface as part of its structure, enabling it to bind directly to the standard interfacing structure. On the bottom is a product-binding molecule, which would be eliminated if the product is built directly on the substrate.

**Positional control
of chemical reactions**

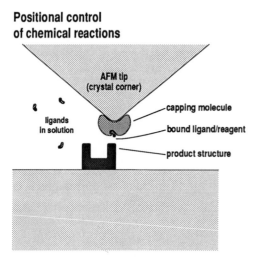

Figure 6.2
Schematic of a molecular manipulator tip.

A device with these characteristics would give positional control of chemical reactions, thereby overcoming the chief obstacle to synthesizing larger structures. This obstacle is reaction specificity. Larger structures have larger surface areas and thus (all else being equal) more sites where a reaction can occur. It is correspondingly more difficult to force a reagent to select a particular site and therefore more difficult to get control of the product structure. Chemists have been clever in working around this limitation. Nonetheless, a direct, flexible technique for determining where a reaction will occur on a molecule would be a fundamental advance. Positional synthesis of this sort would permit the construction of large molecules with complex bonding. The largest molecules made today (with specific control) are linear chains of protein and DNA, which have been synthesized using the trick that only one reactive site is available at a time: the one at the end of the growing chain. Positional synthesis will overcome this constraint.

A more specific concept is shown in figure 6.3. This shows the mobile structure as the tip of an AFM (the positioning mechanism is the entire, macroscopic AFM apparatus). In this concept, the tip is the corner of a crystal. This corner need not be an atomically sharp structure, but if it is an atomically defined structure, then one can perhaps design a mole-

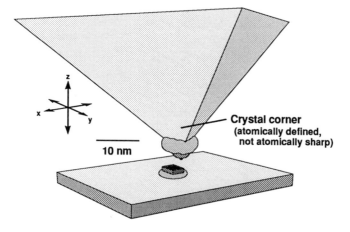

**Figure 6.3**
Schematic AFM-based molecular manipulator.

cule that binds to it specifically. The result would be a more readily controllable tip structure, because once one surface of a molecule is designed to bind the crystal corner, the other surface can be modified to provide an interface to transient tip molecules or reagents.[7]

The great sacrifice in these schemes, of course, is that such devices could make only one molecule at a time. One molecule can be valuable, however, if it yields either useful information or can serve as a tool for making more.

Another approach to making a molecular manipulator with similar capabilities (but greater potential parallelism) would be to build a system in which the positioning mechanism itself is a self-assembled complex of macromolecules. This is a more ambitious goal, requiring additional molecular engineering. To suggest one mechanism for actuating and controlling such a system, it might be built of folded polymer molecules that can reversibly bind small solute molecules, where binding and releasing a solute molecule causes a conformational change in the folded polymer. Each degree of freedom in the molecular-manipulator structure could be associated with one type of binding site and a corresponding solute molecule. By changing the composition of the solution, we could change the shape of the molecular aggregate. With a suitable set of

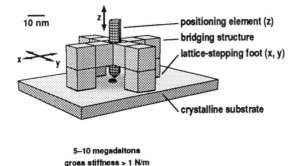

**Figure 6.4**
Schematic positioning system for a molecular-scale molecular manipulator.

available states, we could design a system that would step along a regular surface in defined and controllable increments.

A crude sketch of what such a device might look like is shown in figure 6.4. It is built of stiff aggregates of protein molecules. Each aggregate is depicted schematically as a cube and is formed from a number of protein molecules (which are not cubical). These would form "feet" that could step across a lattice in the x and y directions by cyclical changes in the composition of the ambient solution. A similar mechanism would move the central rod in the z direction. As before, there is a region containing a tip, a reagent binding site, and a bound product, but now the surrounding mechanism is measured in tens of nanometers, rather than tens of centimeters.

A comparison of these schemes for a molecular manipulator suggests the following conclusions. The AFM-based systems appear to be simpler and faster. The tip can be repositioned at electromechanical speeds, which are higher than the rates at which solutions can be cyclically changed. With fewer molecules to be engineered, design and implementation are less difficult. An AFM-based system is also self-instrumenting; given the right tip structure, it can operate as a standard AFM giving feedback on the results of synthetic operations. I have been concerned, however, that there might be a variety of mechanical problems associated with the flexibility of the tip molecules, including the bending stiffness of the arm in the vertical dimension, bending stiffness in the transverse directions, and forces associated with surface tension on tips immersed in a solution. After considering the contributions to this volume, these problems do not seem so large as they once had. Nevertheless, a host of molecular engineering and AFM engineering issues remain to be dealt with. Clearly, the AFM-based system is not simple—it's just simpler than presently known alternatives.

The great virtue of purely molecular systems is their potential for working in parallel and in great numbers. If their components are proteins, or proteinlike molecules, made by Merrifield solid phase synthesis, they can be made in gram quantities. The mass per device is small enough that $10^{15}$ such devices can work simultaneously in a convenient volume. They would be working without the direct feedback possible with the AFM-based systems, and with a finite error rate, but they could yield macroscopic quantities of product. The product molecules could be made to self-assemble, because they could be present in adequate concentrations in a solution. The opportunity to make on the order of $10^{15}$ product structures at a time instead of one at a time may motivate an effort in this direction, despite the challenge of greater molecular complexity.

In both of these approaches, the tips and their associated physical phenomena would be similar, hence both classes of molecular manipulator would have similar synthetic capabilities. It might be desirable to develop both, using AFM-based systems to perform prototype reactions and scientific studies, and later using purely molecular systems to produce useful quantities of product.

## Advanced Assemblers

The previous discussion is a systems engineer's overview of a number of capabilities and how they might fit together to make something like an assembler—though with sharply limited capabilities. By greatly extending our ability to synthesize complex structures, however, molecular manipulators should enable the construction of better molecular-manipulating assemblers. For example, these devices will enable researchers to build complex molecular objects that are not folded polymers but structures put together in a "tinker toy" fashion. These structures can be relatively rigid, with a broad range of physical properties, and can be used to make an even more capable generation of devices. Because these properties will make the construction and design processes easier, turnaround times in the design and debugging cycle will become shorter.

Progress along these lines should lead to a growing family of molecular machines of broad capabilities.

## Exploratory Engineering

I have defined nanotechnology as giving us "thorough and inexpensive control of the structure of matter." This is a capability that does not yet exist. How can one study such a field? One appropriate methodology is what I have termed *exploratory engineering*. In exploratory engineering, the key to establishing more than one might expect is to sacrifice many objectives that one might be expected to pursue.

In standard engineering, a major constraint is that products be designed in such a way that they can be built in the near future and be competitive in the marketplace or on the battlefield. Standard engineering thus attacks much harder problems than exploratory engineering, which merely aims to build a sound case for the feasibility of a class of devices—not to build them, not to beat the competition, but simply to make a sound argument. This permits trade-offs in methodology, argumentation, and analysis. One can sacrifice objectives that would be essential if one were trying to build a product and have it be competitive in exchange for a more direct, defensible case for feasibility. A fundamental difference is that in conventional engineering, one has an incentive to stretch today's tools to their limits to be competitive, while in exploratory engineering the goal is to find very ordinary uses for tomorrow's tools. Typically, the results of exploratory engineering are sophisticated in that they depend on sophisticated tools (that is, the results assume a powerful manufacturing technology base), but they are crude in the sense that by the time we learn how to build such things, the original approach would be outdated. An engineer in the future who dusts off our old designs is likely to say, "Well, I suppose that if we correct a few details, we *could* build something like this, but why would we when we now know how to build something ten times better."

Therefore, the exploratory designs that I am about to present cannot be built today, and it seems that no one is likely to want to build them tomorrow, yet these design are useful for understanding what is possible, for setting lower bounds on future capabilities. I believe that if we are to make sound decisions about the future, we must have at least a rough

A

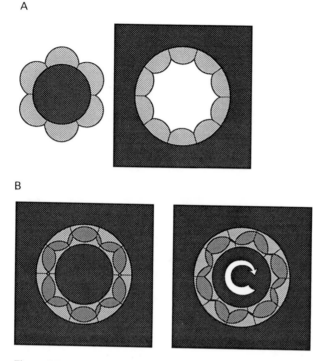

B

**Figure 6.5**
Steric-repulsion bearing model: inner and outer rings (A), rings in two different rotational positions (B).

idea of what we will be able to do in that future. This has been my primary motivation in pursuing such an unusual line of inquiry.

## Component Structures of Assemblers

Figure 6.5 illustrates an idea that is likely to be used in assemblers because it is basic and simple: a steric-repulsion bearing. This is a simple physicist's model of a bearing that centers a shaft in a sleeve using steric repulsion forces between the atoms on the inner and outer rings. The question is, can you actually rotate this bumpy shaft inside this bumpy sleeve? The overlaps of the atoms (exaggerated in figure 6.5) are similar in two different positions of the rotating shaft, but the exact pattern is different. (See appendix A, Machines of Inner Space, for a more physically realistic picture, with two cylinders instead of two rings.)

The force that must be applied tangentially at the edge of the bearing

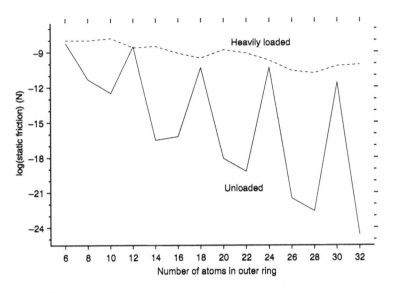

**Figure 6.6**
Static friction in steric-repulsion bearing.

to overcome static friction (that is, to make the shaft go over the energetic bumps) can be analyzed in terms of spatial Fourier transforms of the pairwise interatomic potential, which have little amplitude at sufficiently high frequencies, and the high-order symmetry of the potential energy function under relative rotation of the shaft and sleeve, which makes small displacements a symmetry operation. The results are shown in figure 6.6 for a case intended to represent unrealistically high values of friction (i.e., for a small number of atoms in the central ring and for a high pressure, both of which are adverse cases).[8] The analysis reveals that as the number of atoms in the outer ring increases, the force required to overcome static friction decreases steeply from $10^{-9}$ newtons (a fraction of the force required to break a single covalent bond and comparable to the force exerted by an AFM tip in scanning a surface) to $10^{-24}$ newtons, which is quite negligible in such a context. The peaks in the plot for the unloaded system reflect combinations of greater and lesser commensurability between the number of atoms in the inner and outer rings (points that would have caused even higher forces have been omitted). Thus, low static friction requires suitable choices of the numbers of atoms.

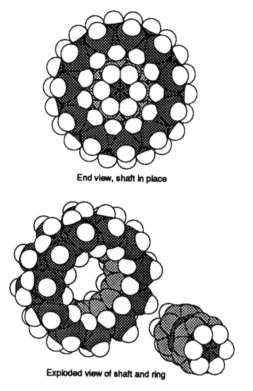

**End view, shaft in place**

**Exploded view of shaft and ring**

Figure 6.7
A bearing design, shown with the shaft in the sleeve (above) and in an exploded view (below).

This result, however, depends strongly on the axial symmetry of the bearing. If the central element is displaced under load (the upper, dotted line in figure 6.6), the symmetry decreases and the forces are initially quite high and move downward only slowly. An unrealistically high load has been assumed for this analysis to emphasize friction, but even so, with more than 26 atoms in the outer ring, friction falls to values that will be low for many purposes.

Figure 6.7 shows an alternative bearing design (with the shaft in the sleeve, and in an exploded view. Note that the shaft has sixfold symmetry, while the sleeve has elevenfold symmetry; these numbers are relatively prime. Rotational energy barriers are less than 0.001 kT at room temperature according to the MM2 molecular mechanics model.[9]

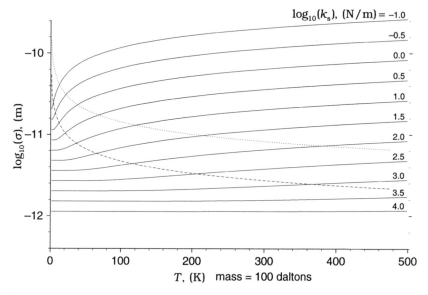

**Figure 6.8**
Root mean square displacements for harmonic oscillators as a function of temperature (oscillator mass on the order of benzene).

### Positional Uncertainty in Nanomechanical Systems

A fundamental issue in nanomechanical systems is positional uncertainty due to the combined effects of quantum uncertainty and thermal excitation. An exact quantum mechanical treatment for harmonic oscillators shows that mean square positional uncertainty is given by the expression

$$\sigma^2 = \frac{\hbar\omega}{k_s}\left(\frac{1}{2} + \frac{1}{e^{\frac{\hbar\omega}{kT}} - 1}\right)$$

presented in terms of the angular frequency of the oscillator, $\omega$, and the stiffness of the restoring force, $k_s$. The classical approximation

$$\sigma^2_{\text{class}} = \frac{kT}{k_s}$$

yields a considerably simpler expression. (Both models yield a Gaussian probability density function.)

As shown in figure 6.8, for a given mass (which matters in the quantum mechanical case but not in the classical case), one can plot the rms deviation in position as a function of temperature for a variety of spring constants. (For a single chemical bond, the logarithm of a typical spring constant is between 2.0 and 3.0; for a typical nonbonded interaction between two atoms in the absence of compressive load, it is typically between $-1.0$ and $0.0$; for bond-angle bending interactions, it is typically between $1.0$ and $1.5$). Above the dotted line, the classical model is accurate to within 1%; above the dashed line, to within 10%.

The analysis becomes more complicated for extended objects. The expression

$$\sigma^2 \frac{\sqrt{E_l \rho_l}}{\hbar}$$

$$= \frac{2}{2N+1} \sum_{n=0}^{N-1} \frac{\sin^2\left(\frac{2n+1}{2N+1}\pi N\right)}{\sin\left(\frac{2n+1}{2N+1}\frac{\pi}{2}\right)} \left( \frac{1}{2} + \frac{1}{e^{\left(\frac{\hbar\omega_0}{kT}\right)\frac{4}{\pi}N\sin\left(\frac{2n+1}{2N+1}\frac{\pi}{2}\right)} - 1} \right)$$

gives a dimensionless measure of the longitudinal positional variance at the end of a rod modeled as a series of springs and masses, expressed as a sum over modes in a system characterized by a mass per unit length $\rho_l$, linear modulus $E_l$, and number of atoms $n$. The classical limit, expressed in terms of the length $l$

$$\sigma_{class}^2 = kT \frac{l}{E_l}$$

is again simple.

The ratio of the quantum to the classical positional variance (the mean square displacement) as a function of temperature is shown in figure 6.9. The properties of the rod are roughly comparable to those of diamond (but use round figures resulting from the origin of this graph, which is taken from a series of evenly spaced slices through this parameter space). Stiff structures exhibit greater quantum effects; nonetheless, for relatively large nanoscale structures (100 nm and larger) the classical result provides an excellent approximation. This family of curves assumes a speed of sound of 20 km/s (a high value which magnifies quantum effects) and interatomic spacing of 0.1 nm (a low value, which also magnifies quantum effects).

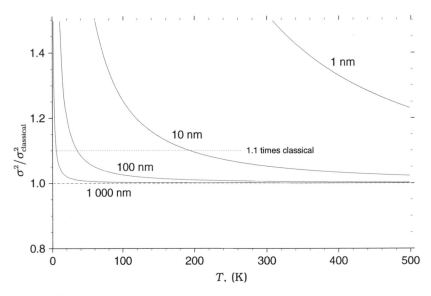

**Figure 6.9**
Ratio of the exact quantum result to the classical limit for the positional variance of elastic rods as a function of temperature.

### Designing an Assembler Arm

These results are useful in understanding what can be done with molecular assemblers, where positional control is crucial. To simplify the analysis for exploratory engineering purposes, it is desirable that the assembler arm be able to position a reactive moiety with extreme reliability, including in the analysis the flexibility of the reactive moiety itself. Reliability means few misplacements, and the probability of a misplacement depends on the distance between reactive sites on the product structure; a typical example might be the 0.25 nm separation between equivalent sites on the (111) diamond surface. If we demand that the probability of an error be $< 10^{-12}$, then 0.25 nm must be about eight standard deviations. Achieving this goal requires a system compliance (reciprocal stiffness) that is $\leq 0.23$ meters/newton at 300 K.

To design an arm to be as stiff as possible, it is advantageous to use a device called a harmonic drive to drive the joints. This is used in macroscopic robotics; on a microscale, its kinematics can be described as making two surfaces move past one another by driving the motions of dislocations (which is not how macroengineers usually think about it).

A schematic of such an arm is shown in figure 6.10. The device is stubby and about 100 nm long, with thick walls of diamond or diamondlike carbon. The upper end has a socket for holding reactive devices and the length of the arm contains a series of joints that give six degrees of freedom control at the end. Overall, its structure contains several million atoms. This is a general-purpose device, making it useful for thought experiments and for setting lower bounds on what could be done. The joints contain steric repulsion bearings like those discussed earlier, but with interlocking grooves to provide good shear stiffness perpendicular to the direction of sliding in the interface. Power and control in this design are supplied by turning a set of slim, flexible shafts threaded along the axis of the arm. Although this arm is somewhat arthritic because of its thick walls and so forth, it has a modest range of motion of about 105 nm$^3$.

A compliance budget for a mechanism of this sort is shown in table 6.1, assuming that the product itself is rather stiff. The compliances are chosen to sum to the allowable limit mentioned above, and the contributions to compliance due to the arm itself are small enough that the majority of the compliance budget ($> 60\%$) can be allocated to the reactive moiety, that is, to the bending of the small set of atoms most intimately involved in the chemical reaction that is the object of the exercise. This result is attractive because it is easy to select reactive moieties that have large compliances. It seems possible to relax some of our earlier constraints and work with higher compliances, but as an exercise in exploring simple systems, it is interesting that it seems one can meet this tight constraint on positional accuracy.

### Advantages of Advanced Assemblers

What does an advanced assembler buy you in terms of making things? In contrast to solution chemistry, it could replace diffusive transport of materials with mechanical conveyance. Initially, this is a cost rather than a benefit, because this demands that one build a mechanism to do the conveyance. The motivation for providing a conveyance mechanism, however, is to completely replace diffusive encounters by positioning, thus preventing any encounter of a reactive moiety with an undesired site.

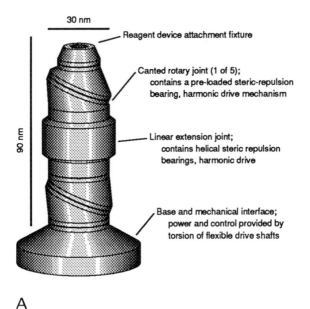

30 nm

Reagent device attachment fixture

Canted rotary joint (1 of 5);
    contains a pre-loaded steric-repulsion
    bearing, harmonic drive mechanism

90 nm

Linear extension joint;
    contains helical steric repulsion
    bearings, harmonic drive

Base and mechanical interface;
    power and control provided by
    torsion of flexible drive shafts

A

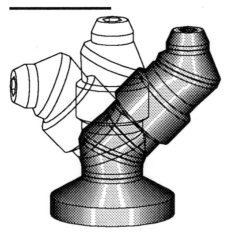

50 nm

B

**Figure 6.10**
Schematic of a rigid nanopositioning mechanism (A); range of motion (B).

**Table 6.1** Positioning mechanism compliance budget

|  | meters/newton |
|---|---|
| Bending in tubular structures | 0.03 |
| Torsion in harmonic drives, shafts | 0.02 |
| Shear in steric repulsion bearings | 0.03 |
| Elasticity near reaction site | 0.01 |
| Bending in reactive moiety (balance) | 0.14 |
| Total (by construction) | 0.23 |

The elimination of reaction at undesired sites allows one to use highly reactive molecules without losing selectivity. My current work focuses on free radical reactions occurring in a vacuum; that is, on using abstraction and addition reactions to build up complex structures.

With less reactive molecules, catalysis by tailoring reaction environments becomes important. Conventional chemistry uses control of solvents, but control of the reaction environment using a positioning mechanism (and the molecular tools attached to it) will allow the use of enzymelike catalysis to drive reactions.

Finally, instead of making do with tactics for achieving selective reactions that are based entirely upon the local structure of the reactants—which is a very complex game to play—it will be possible to gain complete specificity based upon direct positional control. These are the central advantages of positional chemistry and its applications to what can be termed *mechanosynthesis*.

## Implications of Nanotechnology

These considerations form a conceptual basis for understanding what nanotechnology (i.e., molecular manufacturing) can mean for other systems. Today our technology is strongly limited by our ability to fabricate things. We do not have billion-bit memories in our computers today—not because no one could design one, but merely because no one can design one that can be manufactured today. I would argue that with assemblers we will be able to manufacture most stable structures. There are some exceptions, but it is more useful to begin with this general conclusion and watch for exceptions than it is to begin by enumerating specific structures that it will be possible to make. With this technology in hand, it seems that the chief remaining limits to what can be done will be physical law (what arrangements of atoms are stable and will do interesting things) and design capabilities (what we are clever enough to design, out of the wider range of physical possibilities). Both of these limits are substantial constraints, and there will be much that can be imagined but remains impossible to construct.

### Consequences for Computation

During the middle of the last century Charles Babbage designed an "analytical engine" to compute with rods and gears of metal. Mechanical computing technology improves as scales shrink, so I have given some attention to what can be done with molecular mechanical computers.[10,11] Some of the conclusions from this work are briefly sketched here.

The technology illustrated here can make thermodynamically reversible NAND gates (figure 6.11). Because these devices do not forget their inputs until they are reset, reversible NAND gates are consistent with the conclusions presented by Norman Margolus in chapter 9. The device physics is simple: the displacement of knobs on one rod can mechanically interfere with the motion of another rod—this is formally analogous to transistors, in which voltage on one conductor can electrically interfere with the flow of current in another.

One can analyze the motions of the rods classically and use classical statistical mechanics to analyze the thermal excitations, the uncertainties in position, and the resulting error rates. Some parameters resulting from this analysis are given in table 6.2. The stated length is not just for a rod embodying a NAND gate with two inputs and one output but rather for a rod that might be used in a programmable logic array, with 16 inputs and 16 outputs. It is long enough (50 nm) for the classical approximation to be rather accurate (see figure 6.9). The displacement associated with a switching event is about 1 nm, and the switching time for a whole series of these devices, mapping 16 inputs into 16 outputs, is 50 ps; the peak speed is in the range of macroscopically familiar speeds, tens of meters per second. The time for the motion is about 17 times the acoustic transit time along the rod, and the peak speed is about 0.002 times the acoustic speed, so that it behaves as a fairly rigid body.

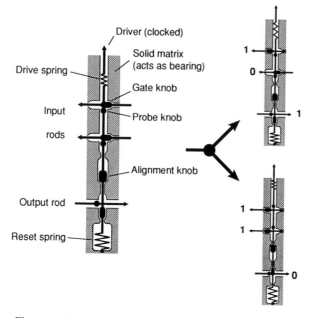

**Figure 6.11**
Abstract diagram of NAND gate operations: components diagrammed on left, two logical states diagrammed on right (immobile = zero, mobile = one).

**Table 6.2**   Logic-rod motion parameters

| | |
|---|---|
| Length | 50 nm |
| Displacement | 0.84 nm |
| Mass | $1.6 \times 10^{-23}$ kg |
| Force | $3.4 \times 10^{-11}$ N |
| Acceleration | $2.1 \times 10^{12}$ m/s$^2$ |
| Travel time | 50 ps |
| Peak speed | 35 m/s |
| Travel time/acoustic time | 17 |
| Peak speed/acoustic speed | 0.002 |
| Peak kinetic energy | $9.8 \times 10^{-21}$ J = 2.4 kT (300 K) |

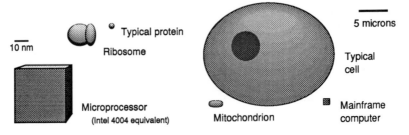

**Figure 6.12**
Scale of a processor similar to an Intel 4004 (A); scale of a nanomechanical computer equivalent to a recent mainframe or a current workstation (B).

Owing to fundamental scaling principles, accelerations in these systems tend to be extremely large, even when forces and stresses are low. The kinetic energy of the rod is modest and is largely recoverable, since it is coupled to an overall drive system (which includes a flywheel) and only weakly coupled to dissipative, vibrational modes. These losses decrease if the clock speed is reduced.

Jumping over a range of details and subsystems, the scale of these nanomechanical gates can be used to estimate the scale of nanomechanical computers. On this basis, we conclude that one could package a nano-mechanical system of the same complexity as an Intel 4004 micropro-cessor (a 4-bit processor, the first that saw any commercial use) into a small box as shown in figure 6.12a. This volume includes no provision for memory, but if we include a comparable space for memory, the resulting volume could hold a minimal computer system. A machine with a capacity equivalent to a mainframe computer of a few years ago, or a workstation of today, could fit inside a cubic micron (figure 6.12b). This provides sufficient volume for a powerful CPU, many megabytes of RAM, and several hundred Mbytes of fast tape memory. Such a machine would still be small compared to a typical mammalian cell, which suggests possible biological and medical applications, perhaps involving computational processes taking data from molecular sensors and using molecular effectors to modify biological structures.

There are many possibilities in this direction, and these have excited a number of imaginative artists. Often these pictures include fanciful details. A painting that appeared in *Scientific American* depicts a device

Table 6.3 Macroscopic system parameters

| Volume | $1 \text{ cm}^3$ |
|---|---|
| CPUs | $10^{12}$ |
| Gate operations | $10^{24} \text{ s}^{-1}$ |
| Power dissipation | 7000 W |
| Cooling water flow rate | $70 \text{ cm}^3/\text{s}$ |
| Cooling water delta T | 25 K |

that could have a mainframe computer in its belly cruising down a capillary with bright red blood cells and chomping on an entirely hypothetical fat deposit in a capillary.[12]

It is of some interest to consider what such computers could do if combined in macroscopic quantities (table 6.3). A cubic centimeter device, with a high-pressure, capillary-style cooling system, could provide some 200 VAX-millennia between screen refreshes, but making effective use of such a machine will require great skill in programming parallel machines since it would have a trillion processors.

**Molecular Assembly Processes**

This computer would be a macroscopic system, a whole cubic centimeter with most of its atoms in the right place (fault tolerance is of course required in such a system because radiation damage, errors in assembly, and statistical extremes in thermal excitation all result in defects). How could such a structure be built? Advanced nanopositioning mechanisms are roughly a million times smaller than conventional macroscopic robot arms, hence scaling laws suggest that they should be a million times faster. They have an additional advantage in that the components that they manipulate—atoms and molecules—are fundamental, inexpensive, and abundant. All of this makes it relatively easy to design an assembler-based device capable of self-replication from simple raw materials. Easy, that is, in comparison to macroscopic mechanical systems. An exercise from the nanotechnology class I taught at Stanford indicated that you could design a replicator composed of a billion atoms, with several nanopositioning arms, that could replicate itself in about 1000 seconds, given raw material inputs of suitable quality. This suggests that molecular manufacturing will eventually be inexpensive because both the materials and the automated manufacturing devices will be inexpensive.

If so, one would expect a large range of products: computers, medical applications, strong materials like diamond-fiber composites, and other diverse applications in the home, in the military, and elsewhere. Large-scale implications are accordingly possible, raising serious public policy issues.

## Paths to Nanotechnology

Clearly we are still far from advanced assemblers and a direct encounter with their implications. Some ambitious intermediate goals of molecular systems engineering—before molecular manufacturing and related nanotechnologies—include the following:

• Routine production of tailored catalysts for specific reactions.
• Terabyte chip memory devices using self-assembled arrays of molecular components.
• Post-integrated-circuit electronics technology.
• Molecular research exploiting positional reaction control.

The first three of these goals are spin-offs and are not on the critical path to developing molecular manufacturing. For example, one need not replace integrated circuits before developing positional control of chemical reactions. The last goal, however, is intimately involved with the developments on the pathway to nanotechnology.

Research areas seem likely to pace the development of molecular manufacturing including:

• Folding-polymer computer aided design (CAD) systems.
• Computational molecular modeling. (It will often be easier to test molecular systems in simulation first.)
• Synthetic organic chemistry. (After testing a system computationally, we need the means to construct it.)
• Molecular object characterization. (Because computer simulations may be defective or inadequate, the product itself must be studied in order to correct any errors in design and manufacturing.)

To be effective, molecular systems engineering needs to be infused with a spirit of engineering, which includes a willingness to avoid difficult problems at the component level in order to tackle more challenging problems at the systems level. This in turn involves generating designs that can be separated into many distinct pieces, making possible broader, more interdisciplinary team efforts than have been common in the molecular sciences.

## Conclusion

I would like to emphasize why we should expect that this path will indeed be followed, leading us to advanced capabilities that now seem remote. Why should we expect nanotechnology to emerge in the real world?

• Many paths lead to assemblers. This chapter outlines several broad paths, and there are many options within each path.

• Payoffs will encourage each step. I have presented some intermediate payoffs in molecular systems engineering; other chapters in this volume that present research in molecular electronics and quantum computation suggest others.

• Nanotechnology has vast commercial, medical, and military potential. This combination of motivations usually suffices to motivate the decision makers in the world.

• The world holds many competing companies and governments. It is hard to imagine that in such a world people will not take the steps that will lead to broad, thorough, and (eventually) inexpensive control of the structure of matter.

## Notes

1. H. Simon, "The Architecture of Complexity," *Sciences of the Artificial*, 2nd ed (Cambridge, Massachusetts: MIT Press, 1981).
2. Merrifield synthesis is a method of solid-phase synthesis of polymers devised originally by R. B. Merrifield to synthesize peptides. In this method amino acids are added stepwise to a growing peptide chain that is linked covalently to an insoluble matrix. Synthesis proceeds from the carboxyl to the amino terminus, the opposite of the direction of biological synthesis. Because amino acids contain, in general, several portions active to the chemistry used for polymerization, complex collections of reagents have been devised to alternately protect and then de-protect active chemical groups on the growing chain so that only the proper active group is available for reaction at any given time. Reactive groups on the amino acid side chains are de-protected at the end of the process.

3. K. E. Drexler, "Molecular engineering: An approach to the development of general capabilities for molecular manipulation," *Proc. Natl. Acad. Sci. USA* 78 (1981): 5275–5278.

4 W. DeGrado, L. Regan, S. Ho, "The design of a four-helix bundle protein," *Cold Spring Harbor Symposia on Quantitative Biology,* vol. 2 (1987) 521–526.

5. M. Mutter, "Nature's rules and chemist's tools: a way for creating novel proteins," *Trends in Biochemical Sciences,* 13 (1988): 260–265.

6. K. E. Drexler, "Strategies for the design of protein-like molecules," (unpublished manuscript, 1989).

7. This concept has been developed further and its performance for positional synthesis has been estimated. See K. E. Drexler, "Molecular tip arrays for molecular imaging and nanofabrication," *J. of Vacuum Science and Technology B.,* 9 (1991): 1394–1397. See also K. E. Drexler, J. Foster, "Synthetic tips," *Nature* 343 (1990): 600.

8. K. E. Drexler, "Nanomachinery: Atomically precise gears and bearings," *Proceedings of IEEE Micro Robots and Teleoperators Workshop* (Hyannis, Massachusetts: IEEE, 1987).

9. From the author's postconference work, 1990.

10. K. E. Drexler, "Rod logic and thermal noise in the mechanical nanocomputer," *Molecular Electronic Devices,* ed. F. Carter, R. Siatkowski, H. Wohltjen, (Amsterdam:North Holland, 1988, 39–46).

11. K. E. Drexler, "Molecular machinery and molecular electronic devices," *Molecular Electronic Devices II,* ed. F. Carter (New York:Marcel Dekker, 1987, 549–571).

12. A. K. Dewdney, "Nanotechnology: wherein molecular computers control tiny circulatory submarines,"*Scientific American* 258 (January 1988): 100–103.

13. R. Lerner, A. Tramontano, "Catalytic Antibodies," *Scientific American,* 258 (March 1988): 58–70. See also I. Amato, "Teaching antibodies new tricks: Antibodies that act like enzymes are filling chemists' heads with new visions," *Science News,* 136 (2 September 1989): 152–153,155.

## Discussion

*Tullock:*  You said that there are physical and chemical barriers. There is also a biological barrier. Life in the world is a tough business, particularly for small things. You want to be careful that you are not producing an attractive form of food for very small organisms. Further, many plants and smaller animals release poisons for the purpose of protecting themselves. We have been living with them a long time and have evolved natural resistance to many of them, but it may be necessary to design your devices so that something like pine odor (a natural insecticide) won't disable them.

*Drexler:*   The point here is the potential fragility of systems of molecular machinery. I imagine these devices working in extremely controlled environments.

*Tullock:*   But you're talking about putting these things in people's bloodstreams.

*Drexler:*   There is a distinction to be made here between the internal environment of the mechanism and its interface to the outside world. For advanced systems, I generally assume that the internal environment consists of moving parts within a vacuum. Certain surfaces, from what we know of biology and materials science, can be quite inert. As long as most of the complex, delicate devices are hidden behind a surface tailored to interface with the environment, I think we'll do all right. But we do have to design it that way. That's a crucial point.

*Kantrowitz:*   I have seen you use models that are similar to macroscopic machines, and also pay attention to the replication mechanism of DNA. Another model that I think should be considered is the immune system. This system, when challenged with an antigen, goes through a process of hypermutation that develops cells particularly effective in dealing with the challenge of that particular antigen.

*Drexler:*   In fact the model of the immune system as a mechanism for generating molecular devices that do specific things has been exploited in recent years for the construction of catalysts.[13] The immune system of an organism produces many millions of different protein molecules that the organism effectively tests for binding capabilities. Exploiting this mechanism to generate diversity, and then selecting from that diversity something of use, is another powerful strategy for engineering protein molecules and perhaps pieces of protein molecules that could be incorporated into other designs.

*Audience:*   Building little assembler arms seem hard enough, but what really boggles my mind is trying to achieve independent coordinated control of all those joints.

*Drexler:*   If you think about the complexity of these devices and imagine a substantial volume of space filled with them, the amount of complexity is enormous. But if you think about the control of a single six-degrees-of-freedom nanopositioning device directed by a mechanical nanocomputer in the immediate vicinity, this system is of precisely the same level of complexity as an industrial robot arm and a conventional computer. The complexity is the same; the components are much smaller.

*Audience:* Right now, there are not any computers or industrial robot arms that are self-replicating systems. Do you think that a nanoscale system of that complexity could be a better self-replicating system?

*Drexler:* The argument here is twofold. First there is the issue of parts. If the parts available are complicated enough, then the task is fairly easy. In fact, in Japan there are already robotic factories where the products include those self-same robotic mechanisms. But the inputs are relatively complex, high-quality components. In the molecular domain, we know that self-replication is possible because we have the existence proof of bacteria. The reason bacteria can self-replicate is that, in the molecular domain, there are genuinely identical parts which are abundant and in effect prefabricated. It is as if you had a warehouse full of identical quality-controlled parts ready for assembly. Not having to fabricate the parts is an enormous advantage. There is no need to mine and refine metals, to have machine tools, and so on.

The other fundamental advantage of molecular-scale manufacturing systems is that the time scale is reduced by a factor of $10^6$. Even if prefabricated parts were available in the macroscopic world, the projected 1000-second replication time for nanoscale replicators translates to a 30-year replication time for a similar but macroscopic replicator. How long would it take in the macroscopic world to develop and debug a manufacturing cycle if one iteration of the cycle took 30 years?

*Audience:* During natural replication, mistakes occur in, for example, the incorporation of amino acids into protein. What is the tolerance in nanoreplicating systems for errors?

*Drexler:* To be conservative, this work assumes that if a single atom is out of place in a bearing, computer memory, or whatever, the device does not work. I assume a zero tolerance for error at the level of functional modules. If you want reliable system behavior at a higher level despite modest error rates at the component level, you must design for fault tolerance and redundancy. This is a familiar topic in computer science, aerospace, and all areas of ambitious systems engineering. The question is, can we build reasonably large modules that are likely to be errorfree? One approach is to assemble large modules from smaller modules that have a high probability of being error free. If you can test modules before using them, you can tolerate a substantial error rate for individual molecular assembly operations and still have a low error rate—even zero—in large systems. This is analogous to the way nature discards bad proteins before forming, say, ribosomes.

For advanced systems, I am working to develop manufacturing processes where you can argue that the error rate per operation is of the order of $10^{-12}$ or less. It is important to consider and model all significant sources of error, but you can make a robust argument for building complex systems even with significant error rates in molecular subassembly.

*Audience:*  Would you care to estimate when we might see the first assembler?

*Drexler:*  I don't know how to calculate such a thing. I have been known, when pressed, to say that I think we are talking about a time in the first third of the twenty-first century. This is a large time window if you consider the rates of advance in computer-aided design technology, systems design, and the synthesis of macromolecules—particularly if you consider the synergistic interactions between these areas.

*Audience:*  Thinking about your statement that in the long term physics and design are the primary limits, what are the long-term design constraints on nanocomputers?

*Drexler:*  I am not an expert in digital switching theory, but, regarding design problems, it is my understanding that one can build switching systems, taking many inputs and mapping them to many outputs, with rather desirable scaling properties. The problem is that the computing power required to solve the design problem increases steeply with the scale of the problem (i.e., the problem is NP-complete). Even design problems of moderate size can require more computational capacity than seems available in the volume of the known universe, even with nanocomputers. This is the sort of solid constraint that can be encountered in the design domain. Other than these limitations, there are all the usual limitations of imagination and luck.

*Audience:*  What I meant is a bit different. We know that there are hard computational problems. But you suggest that when you are designing mechanical nanocomputers, that you are doing it conservatively. What design problems would you expect to encounter trying to build a one-cubic-centimeter, trillion processor computer as you described?

*Drexler:* The most serious design problems for such computational systems will likely be software problems. If we had that many processors today, we could map certain algorithms onto them immediately. But for many problems we suspect that we could do something useful with them if we knew how, but we don't. It will require cleverness and insight to figure out how to exploit so many processors. There are probably clever architectures that will take a long time to invent. In the meantime, it will be possible to build relatively simple parallel machines with our present knowledge.

*Audience:* Are there any near-term steps toward the development of nanotechnology that are narrowed down to only a few choices, and all of these choices are very expensive?

*Drexler:* I don't think so. By the standards of large-scale enterprises, the costs of doing research in chemistry and related fields are small. Also, there are many alternative approaches for solving each problem.

*Audience:* Dr. Birge stated that $10^5$ molecules of rhodopsin were needed to get one bit without error. Do you agree that this many molecules are needed in an assembly to get reliable computation? Can you imagine building an assembly of $10^5$ molecules in an orderly fashion?

*Drexler:* You are raising two questions here. One is the feasibility of making a structure of $10^5$ molecules, by chemical synthesis and self-assembly of folded polymers. The other is how many molecules are needed for a reliable computational element. With regard to the first, I don't know whether a structure of that scale is feasible using those techniques. If not, then one postpones the attempt until one has a better controlled, more reliable mechanism for construction, such as a molecular manipulator or a full-fledged assembler. Regarding how many molecules are needed to do a computation, this depends on the physical nature of the computational operation. If you are doing fast optical switching, the number of molecules needed may be quite large. We do know, however, from examples in molecular genetics, and from a detailed analysis of rod-logic systems, that a single molecular-scale component can perform reliable computational operations by chemical and mechanical means.

# II

## Related Technologies

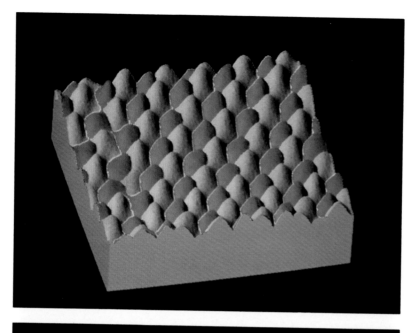

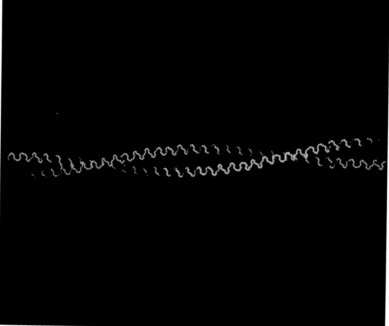

**Plate 1**
Indium phosphide surface. (Courtesy Jun Nogami, Stanford University.)

**Plate 2**
Polypeptide backbone representation of an α-helical coiled coil.

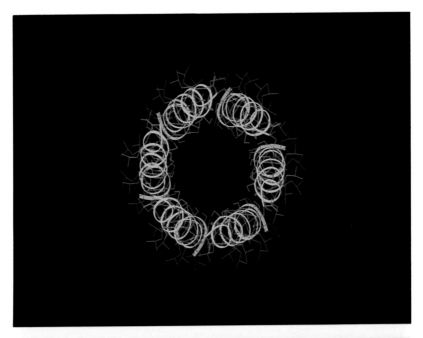

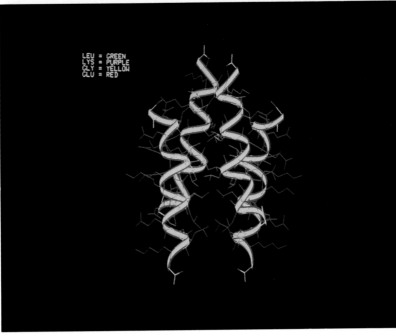

**Plate 3**
Polypeptide backbone representation of an ion channel.

**Plate 4**
Polypeptide backbone representation of a 4-helix bundle protein.

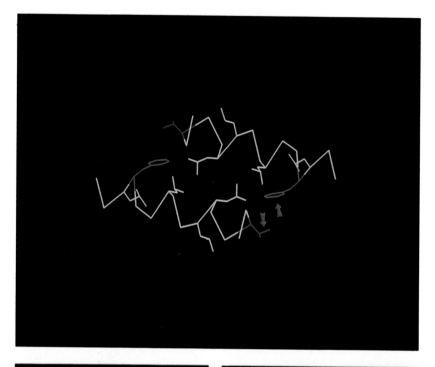

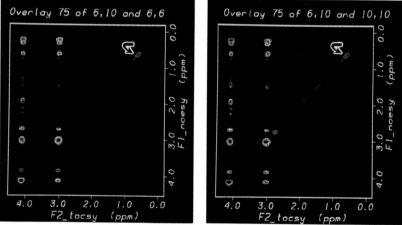

**Plate 5**
*Top*, interaction expected between valine and phenylalanine in Val₃Phe₁₃, a
"mutant" α₁B peptide.

**Plate 6**
*Bottom left*, overlay NOESY-TOCSY of 50:50 mixture of Val₃Phe₁₃ (H-Leu₆)
plus Val₃Phe₁₃ (H-Leu₁₀) with NOESY-TOCSY of Val₃Phe₁₃ (H-Leu₆).

**Plate 7**
*Bottom right*, overlay NOESY-TOCSY of 50:50 mixture of Val₃Phe₁₃ (H-Leu₆)
plus Val₃Phe₁₃ (H-Leu₁₀) with NOESY-TOCSY of Val₃Phe₁₃ (H-Leu₁₀).

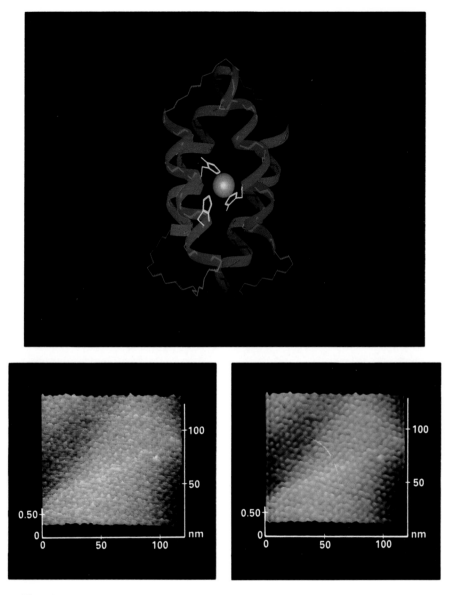

**Plate 8**
*Top*, molecular model of a subunit of H3α₂.

**Plate 9**
*Bottom left*, packing of photoactive enzymes: original STM data.

**Plate 10**
*Bottom right*, packing of photoactive enzymes: Fourier transformed data.

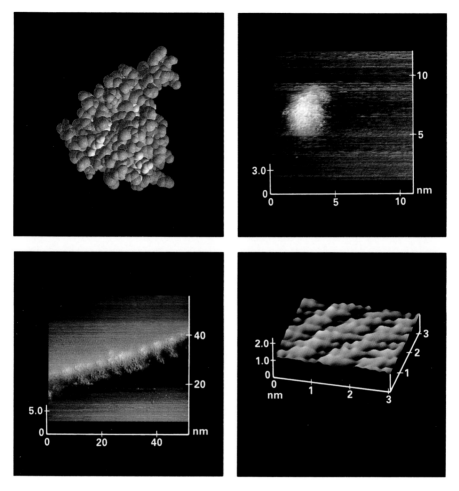

**Plate 11**
*Top left*, computer graphic visualization of the structure of cytochrome c3 as determined by X-ray diffraction.

**Plate 12**
*Top right*, STM of cytochrome c3 adsorbed to a surface.

**Plate 13**
*Bottom left*, STM of bacteriorhodopsin molecules adsorbed to a graphite substrate.

**Plate 14**
*Bottom right*, STM of dicyano vinyl anisole.

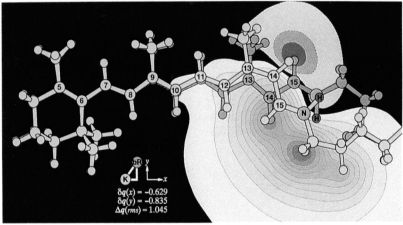

$$\delta q(x) = -0.629$$
$$\delta q(y) = -0.835$$
$$\Delta q(rms) = 1.045$$

**Plate 15**
Photomicrograph of the bacteria *Halobacterium halobium*, in a salt solution, from which bacteriorhodopsin is isolated.

**Plate 16**
A model of the primary photochemical event in bacteriorhodopsin.

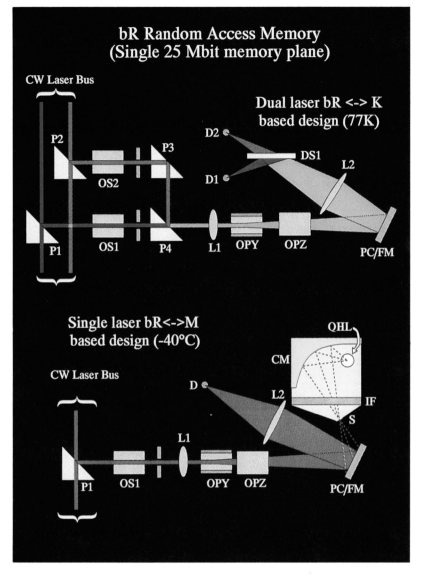

Plate 17
Schematic diagram of the principal components of a single memory plane of a bacteriorhodopsin-based optical random access memory.

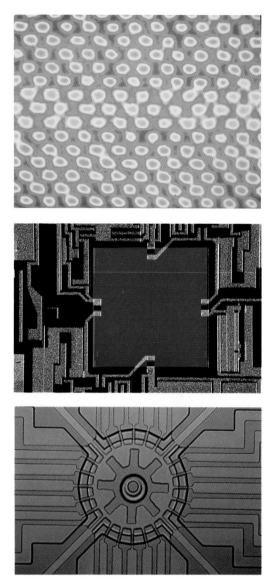

**Plate 18**
*Top,* TEM of GaInAs/InP quantum well created with MBE. (Courtesy M. B. Panish, AT&T Bell Laboratories.)

**Plate 19**
*Center,* transmission micrograph of micromachined silicon pressure sensor.

**Plate 20**
*Bottom,* silicon micromotor. (Courtesy Y. Tai, R. Muller, Berkeley Sensor and Actuator Center, University of California, Berkeley.)

# 7

## Molecular Electronics

Robert Birge

Molecular electronics is an emerging field that lies at the interface of molecular physics, electrical engineering, optical engineering, and solid-state science. It involves the encoding, manipulation, and retrieval of information at a macromolecular level as opposed to current techniques, fast approaching their practical limits, in which these functions are accomplished via miniaturization of bulk devices, such as integrated circuits. This field not only represents the final technological stage in the miniaturization of computer circuitry but it also provides promising new methodologies for high-speed signal processing, optical data storage, and content-addressable memory devices. Biomolecular electronics is a subfield of molecular electronics that involves the use of native as well as modified biological molecules (chromophores, proteins, etc.) in place of the organic molecules synthesized in the laboratory. Because natural selection processes have often solved problems of a similar nature to those that must be solved in harnessing organic compounds, biomolecular electronics has dominated the early stages of this research effort. This chapter overviews our recent research at the Center for Molecular Electronics at Syracuse University in the area of molecular and biomolecular electronics.

This overview of molecular electronics includes both prototype and long-range basic research projects, and it is important to note at the outset that a majority of the devices under investigation are designed to test concepts rather than test commercial viability. An emphasis is placed on discussing molecular electronic devices that are relevant to computer architecture.

Molecular electronics is a global term that describes the use of organic

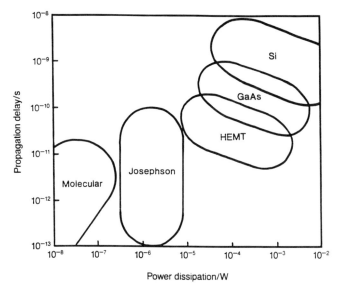

**Figure 7.1**
Canonical propagation delays for various types of logic gates and power dissipation for several device technologies.

components or materials as active elements in electronic devices or environments that until recently were considered the exclusive domain of metallic conductor and inorganic semiconductor materials. Examples include conducting polymers, molecular rectifiers, nonlinear organic compounds, protein-based digital and associative optical memories, and optically coupled molecular gates.

Figure 7.1 illustrates the propagation delay through canonical electronic gates of various construction, displayed on a logarithmic scale. Exemplifying the current apex of silicon (Si) devices, a typical silicon-based supercomputer now operates at about a 10 nsec cycle time. Replacing silicon with gallium arsenide (GaAs) is expected to give one order of magnitude greater speed, and high electron mobility technology transistors (HEMT),[1] such as gallium arsenide (GaAs), should yield another order of magnitude improvement. About the best cycle time that can be expected with large-scale integrated HEMT devices is about 10 psec (100 GHz)—although individual HEMT transistors have been pushed beyond this rate. Josephson device technology allows two orders of magnitude improvement over current silicon technology.[2] However, the implementation of this technology is complicated by the problem that Josephson gates have only two junctions and the amplification process that is required for fan-out implementations is difficult to achieve. Thus, design changes that are introduced in one portion of the overall design will have an impact on the entire system. IBM, after investing heavily in research on Josephson-based computational networks, scrapped a majority of the effort in recognition of the above problems. Simultaneously, a small group of researchers were given permission to investigate the potential of molecular electronics, a quiet but significant admission that this technology might offer potential for future computer architectures.

Molecular electronics has the distinct advantage of providing the high transition rates and efficiency that is characteristic of the small dimensions of the molecular regime. Of equal importance are the new types of architectures that molecular technology can provide.[3]

Although the potential of molecular electronics is sufficient to drive enthusiasm, there is another aspect of this research effort that deserves recognition. If we project the dimensions of semiconductor features into the future, the current trends suggest that minimum feature sizes will approach the nanoscale regime within the next decade. Thus the quantum electronic and reliability problems that we are dealing with today in trying to implement molecular technology are problems that the semiconductor scientists will be dealing with in the near future. As Federico Capasso describes, in chapter 8, quantum mechanical effects can provide a significant advantage in implementing unique and efficient architectures. However, quantum mechanics is also an added complexity that requires sophisticated computer-aided design capabilities and nanoscale accuracy in implementation.

Comparing molecular with semiconductor electronics reveals four principal advantages and four principal disadvantages (table 7.1). Molecular electronic devices are 100 to 1000 times faster than current silicon technology and, if properly implemented, more energy efficient. Some might object that molecular devices are inherently unreliable, and thus reliability problems should be considered an inherent disadvantage. However, if molecular devices are designed to include ensemble averaging, they can be designed to achieve very high reliabilities. Because a number of architectures can be implemented more easily based on mo-

**Table 7.1**    Advantages and Disadvantages of Molecular Electronics

| Advantages | Disadvantages |
| --- | --- |
| 1. Speed | 1. Quantum statistical uncertainties |
| 2. Efficiency | 2. Thermal instabilities |
| 3. Controlled reliability | 3. Large-scale synthetic problems |
| 4. Unique design capabilities | 4. External communication problems |

lecular components, there are inherent advantages for certain applications such as content addressable (associative) memories and neural network architectures.

Unfortunately, the problem of quantum statistical uncertainties has been overlooked in a number of articles on molecular electronics. I believe the correct addressing of this issue is intrinsic to the construction of any reliable device based on molecular components. For example, thermal instabilities can cause reliability problems because of the breakage of chemical bonds via random thermal motion. A system design that relies upon a single chemical bond for the entire system to function properly will obviously malfunction if that bond breaks. There are many designs in the literature that will not work because a critical bond has a high probability of breaking under the stresses imposed by continuous operation. There are also substantial problems involved in synthesizing the large, complex molecules that are designed to carry out simple logic functions. Our research is held up more by synthetic problems than by any other experimental hurdle. (We will demonstrate later that harnessing complex biological molecules can provide significant help in solving the synthesis problem.)

Finally, we must find efficient and nondestructive methods of communicating with molecules. The technique that we have found to be the most reliable is optical coupling. That is, using light to read, write, or activate molecules. Scientists have worked on activating molecules by light for decades, and the entire field of molecular spectroscopy is applicable to molecular-based photonic and optical computing. The methods and procedures of molecular spectroscopy are well understood, and the development of lasers has made optical coupling a technique that can be implemented in the short term. Of equal importance is the ability to carry out ensemble averaging by using optical coupling to activate or interrogate more than one molecule simultaneously. It is important to recognize that photonic ensemble averaging carries out ensemble averaging without any diminution in the intrinsic speed of the device, because all of the molecules to be activated can be activated with near simultaneity.

## Quantum Uncertainty in Molecular Electronic Devices

The properties of isolated molecules and the behavior of molecules under the influence of electrical fields or optical excitation can only be determined reliably using quantum theory. While classical and semiclassical methods have often been applied with success in modeling such phenomena, these successes should not be used as an argument to ignore quantum mechanics. Quantum theory is particularly important in evaluating the reliability of molecular-scale devices.

Heisenberg's uncertainty principle states that the product of the energy uncertainty and the time uncertainty is equal to or greater than $\hbar/(4\pi)$, where $\hbar$ is Planck's constant ($\hbar \cong 6.6261 \; 10^{-34}$ J s). If we equate the time uncertainty to the aperture time for state assignment, a device operating with a one-picosecond ($10^{-12}$ s) aperture time will introduce a minimum energy uncertainty of about 3 cm$^{-1}$. (More sophisticated calculations yield a value roughly three times larger than this for the expectation value of the energy uncertainty.) In practical terms, these calculations indicate that the separation between molecular states representing the bit states of the device must exceed 10 cm$^{-1}$ for reliable and reproducible state assignment. This energy differential is easily met by a majority of the molecular scale devices proposed in the literature. But if we propose a device operating at an aperture time of 10 femtoseconds ($10^{-14}$ s), the energy uncertainty increases to $\sim 1000$ cm$^{-1}$, and reliable state assignment becomes significantly more difficult. We can therefore conclude that there are quantum-mechanical limitations on the speed of both molecular and semiconductor devices that may well prohibit reliable operation in the femtosecond regime.

Of equal importance is the issue of state assignment. If we use the electronic or conformational states of a molecule to represent a binary bit of information, a key requirement of a molecular-scale computational device is the ability to assign the state of the molecule with high accuracy.

The probability of correctly assigning the state of a single molecule, $p_1$, is never exactly unity. This less-than-perfect assignment capability is due to quantum effects as well as inherent limitations in the state assignment process. The probability of an error in state assignment, $\mathcal{P}_{error}$, is a function of $p_1$ and the number of molecules, $n$, within the ensemble used to represent a single bit of information. $\mathcal{P}_{error}$ can be approximated by the following formula:[4]

$$\mathcal{P}_{error}(n, p_1) \cong -\text{Erf}\left[\frac{(2p_1 + 1)\sqrt{n}}{4\sqrt{2p_1(1 - p_1)}} ; \frac{(2p_1 + 1)\sqrt{n}}{4\sqrt{2p_1(1 - p_1)}}\right], \tag{1}$$

where Erf $[Z_0 ; Z_1]$ is the differential error function defined by

$$\text{Erf}\ [Z_0 ; Z_1] = \text{Erf}\ [Z_1] - \text{Erf}\ [Z_0], \tag{2}$$

where

$$\text{Erf}[Z] = \frac{2}{\pi^{1/2}} \int_0^z \text{Exp}\ (-t^2)\ dt\ . \tag{3}$$

Equation 1 is approximate and neglects error associated with the probability that the number of molecules in the correct conformation can stray from their expectation values based on statistical considerations. Nevertheless it is sufficient to demonstrate the issue of reliability and ensemble size. First, we introduce a reliability parameter, $\xi$, which is related to the probability of error in the measurement of the state of the ensemble (device) by the function, $\mathcal{P}_{error} = 10^{-\xi}$. Note that as $\xi$ increases, reliability increases logarithmically. A value of $\xi = 10$ is considered a minimal requirement for reliability in non-error-correcting digital architectures.

If we assume that the state of a single molecule can be assigned correctly with a probability of 90% ($p_1 = 0.9$), then equation 1 indicates that 95 molecules must collectively represent a single bit to yield $\xi > 10$ [$\mathcal{P}_{error}(95, 0.9) \cong 8 \times 10^{-11}$]. We must recognize that a value of $p_1 = 0.9$ is larger than is normally observed; some examples of reliability analyses for specific molecular-based devices are shown in figure 7.2. In general, ensembles larger than $10^4$ are required for reliability unless fault-tolerant or fault-correcting architectures can be implemented (see below).

The question then arises whether or not we can design a reliable

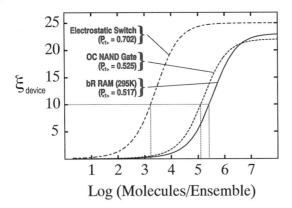

Log (Molecules/Ensemble)

**Figure 7.2**
The effect of ensemble size on the reliability of selected molecular devices represented by plotting $\xi$ as a function of $\log_{10}$ (molecules/ensemble). The reliability parameter, $\xi$, is related to the probability of error in the measurement of the state of the ensemble (device) by the function, $\mathcal{P}_{error} = 10^{-\xi}$. Note that as $\xi$ increases, reliability increases logarithmically. The error probability, $\mathcal{P}_{error}$, derives from quantum statistical error in the measurement of state as well as the probability that the number of molecules in the correct conformation can stray from their expectation values based on statistical considerations.

computational system that uses a single molecule to represent a bit of information. The answer is yes—provided one of two conditions apply.

The first condition is architectural. It is possible to design fault-tolerant architectures that either recover from digital errors or simply operate reliably with occasional error due to analog or analog-type environments. An example of digital error correction is the use of additional bits beyond the number required to represent a number. This approach is common in semiconductor memories, and under most implementations these additional bits provide for single-bit error correction and multiple-bit error detection. Such architectures lower the required value of $\xi$ to values less than 3. An example of analog error tolerance is embodied in many optical computer designs that use holographic and/or Fourier architectures to carry out complex functions.

The second condition is more subtle. It is possible to design molecular architectures that can undergo a state reading process that does not disturb the state of the molecule. For example, an electrostatic switch could be designed that can be "read" without changing the state of the switch. Under these conditions, the variable *n*, which appears in equation

1, can be defined as the number of read operations rather than the ensemble size. Thus our previous example indicating that 95 molecules must be included in the ensemble to achieve reliability can be restated as follows: a single molecule can be used provided we can carry out 95 nondestructive measurements to define the state.

## Optically Coupled NAND Gates

Despite the potential of fault-tolerant architectures, most of the devices we have investigated require ensemble averaging to achieve reliability. The fact that more than one molecule is required to represent a single bit of information does not represent an inherent limitation in the application of molecular electronics. In a majority of cases, ensemble averaging can be used to enhance reliability without a loss of device speed. The most convenient method of implementing ensemble averaging is the use of optical coupling.

In this section I discuss some research done by my students and collaborators on optically coupled NAND gates. The NAND gate is the fundamental logic gate from which all others can be formulated. It is thus appropriate to choose this gate as the initial goal in the development of molecular-based logic. Two designs of optically coupled gates currently under study are shown in figure 7.3. These two designs operate via different activation mechanisms.

I will discuss the key operating principles and design criteria of both devices. First, however, I will provide a brief overview of how optically coupled NAND gates operate. Each gate has two input chromophores and one output chromophore. Two pulsed laser beams are tuned to the electronic transitions of the individual gates so that either input chromophore, or both input chromophores, can be activated. A third continuous-wave (CW) laser beam monitors the output chromophore. The gate "fires" only when both inputs are activated; the absorption band of the output chromophore is shifted in wavelength so that it absorbs the CW laser beam. Absorption of the CW beam is assigned to logic level 0, and thus we have a NAND gate. Two NAND gate designs are under investigation. The first type, shown in figure 7.3a, is based on the use of zwitterionic electron transfer inputs. The second type, shown in figure 7.3b, is based on the use of charge transfer excitation inputs.

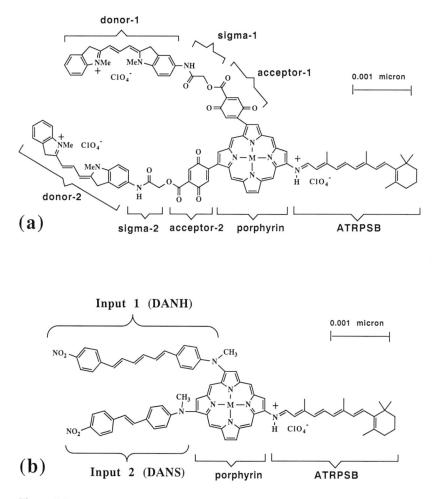

**Figure 7.3**
Two optically coupled molecular electronic NAND gates currently under study.

### Electron Transfer NAND Gate

The gate shown in figure 7.3a operates by using two optically coupled inputs provided by using π-σ-π, donor-sigma-acceptor complexes comprised of a cyanine chromophore (donor) connected by a (-$CH_2$-CO-$CH_2$-) linkage to a quinone chromophore (acceptor). The quinone group is attached to the porphyrin macrocycle (P) which serves as a charge integrator. In the resting state, the cyanine donors have both transferred an electron across the σ barrier to the quinone acceptors ($D^+$-$A^-$). Weak conjugative stabilization of the quinone anion by the porphyrin macrocycle electrostatically stabilizes a retinyl proton-ated Schiff base (ATRPSB) substituted on the porphyrin. Optical exci-tation of the input cyanine-quinone system involves a charge transfer transition that transfers an electron to produce a neutral species (D-A). Because the two inputs have charge transfer bands at slightly different frequencies, selective excitation is possible. If neither input is excited, or only one input is excited, the output chromophore remains electrostat-ically stabilized. However, if both inputs are excited, the stabilization is lost (D-A)$_2$-(P)-(ATRPSB), and the absorption band of the output chromophore (ATRPSB) shifts from ~ 520 nm to ~ 590 nm. The light from a continuous-wave laser tuned to a wavelength in the region of 580–630 nm will experience an increase absorption and will register a logic change from 1 to 0. Hence, the above system acts as an optical NAND gate. The logic reset rates are governed by the tunneling times (~ 1–3 ps) associated with reformation of the zwitterionic states of the inputs.

### Charge Transfer NAND Gate

The gate shown in figure 7.3b operates by replacing the donor-sigma-acceptor complexes with donor-acceptor polyenes. These polyenes are known to have lowest-lying, strongly allowed charge-transfer excitation bands involving HOMO → LUMO excitations. Upon excitation into the lowest-lying excited state, a large shift in electron density occurs with a motion of electron density from the $NO_2$-phenyl group to the phenyl-N($CH_3$)-R end of the polyene. Excitation of an input chromo-phore increases electron density in the porphyrin system and enhances stabilization of the output chromophore. Thus, this gate operates by shifting the absorption band of the output chromophore to the blue, rather than the red, upon "firing." The advantage of this design over the electron transfer NAND design is the simplicity of the system and the ability to tune the input chromophores by using different polyene chain lengths. The disadvantage is that the logic reset rate is a function of the excited state lifetime of the lowest-lying charge transfer bands of the input chromophores and is less easily controlled synthetically.

**Proteins as Molecular Electronic Devices**

I will now discuss proteins as they might be used in molecular electronic applications. There are significant advantages inherent in the use of biological molecules, either in their native form, or modified via chemical or mutagenic methods, as active components in molecular electronic devices. These advantages derive in large part from the natural selection process and the fact that nature has solved—through trial and error— problems of a similar nature to those encountered in harnessing organic molecules to carry out logic, switching, and energy transducing functions. One example of this type of serendipity is the use of the light harvesting protein, *bacteriorhodopsin,* as the photoactive element in optically coupled devices.

During the past twelve years we have investigated the use of two proteins, rhodopsin and bacteriorhodopsin. *Rhodopsin* is the protein in the rod cells of the eye that is responsible for converting light into an optic nerve impulse. While rhodopsin has a number of excellent photonic characteristics, there are two properties that preclude routine application. First, the protein irreversibly bleaches after absorbing a photon of light, a characteristic that is used in living systems to provide for visual light adaptation and high signal-to-noise response to light activation. In biological systems, a series of enzymes is available to regenerate the rhodopsin protein, but these enzymes cannot be maintained in device environments. Second, the only commercial source for this protein is bovine rod outer segments isolated from cow eyes. Because as many as 1000 cow eyes are typically required for a single experiment, this source presents a severe financial limitation. Fortunately, there is a second protein with all of the advantages and neither of the two disadvantages. This protein is called bacteriorhodopsin and can be obtained in large quantities from a bacterium that can be grown in the laboratory.

Plate 15 shows a photomicrograph of the salt-loving bacteria, Halobacterium halobium, from which the protein bacteriorhodopsin is isolated. The photomicrograph has been computer enhanced to more clearly display the bacteria by using a false-color algorithm to show the bacteria in purple and the highly saline solution in green. The rod-shaped bacteria have diameters of $\sim$ 0.7 $\mu$m and lengths of 5 to 10 $\mu$m. The flagella that propel the bacteria are too small to observe under an optical microscope. The bacteria grow the purple membrane containing bacteriorhodopsin when the oxygen concentration decreases to the point that the bacteria are forced to grow the photosynthetic purple membrane in order to sustain growth. The micrograph shows a mutant strain of the bacteria used in our laboratory because it is an "over producer" of the protein. That is, this strain produces much more of the protein than the bacteria actually needs to maintain life. The use of an over producer increases the yield of the protein by twofold.

Bacteriorhodopsin (MW @ 26,000) is the light-transducing protein in the purple membrane of Halobacterium halobium. This halophilic archaibacterium grows in salt marshes where the concentration of NaCl can exceed 4 M, roughly six times higher than seawater ($\sim$ 0.6 M NaCl). The purple membrane, which contains the protein bacteriorhodopsin in a lipid matrix (3:1 protein:lipid), is grown by the bacterium when the concentration of oxygen becomes too low to sustain the generation of ATP via oxidative phosphorylation. Upon the absorption of light, bacteriorhodopsin converts from a dark-adapted state to a light-adapted state. Subsequent absorption of light by the latter generates a photocycle (figure 7.4) that pumps protons across the membrane, with a net transport from the inside (cytoplasmic) to the outside (extracellular) of the membranes. The resulting pH gradient ($\Delta$pH $\sim$ 1) generates a proton-motive force that is used by the bacterium to synthesize ATP from inorganic phosphate and ADP.

At ambient temperatures under low-light conditions, the purple membrane of Halobacterium halobium contains a binary mixture of two proteins, one containing 13-cis retinal and the other containing all-trans retinal. The protein containing 13-cis retinal is known as dark-adapted bacteriorhodopsin, and the protein containing all-trans retinal is known as light-adapted bacteriorhodopsin. The latter protein undergoes the photocycle shown in figure 7.4 and is the only form of the protein that

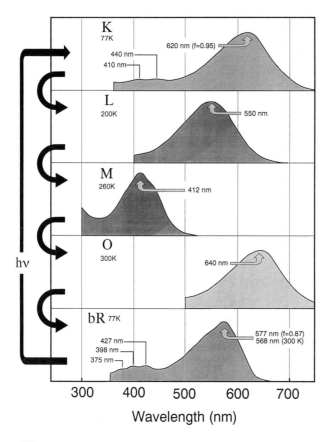

**Figure 7.4**
Photocycle of light-adapted bacteriorhodopsin showing the electronic (one-photon) absorption spectra of selected intermediates in the photocycle. The temperatures used to measure the spectra are indicated underneath the intermediate labels. Band maxima are indicated in nanometers and oscillator strengths (f) of selected $\lambda_{max}$ bands are indicated in parentheses.

pumps protons. Thus, light-adapted bacteriorhodopsin is the biologically active form and in all subsequent discussions we will refer to light-adapted bacteriorhodopsin simply as bacteriorhodopsin, or bR.

The phototransformation of bR to K is the "primary event." The primary photoproduct, K, stores $\sim 12$ kcal mol$^{-1}$ and involves an all-trans to 13-cis photoisomerization of the protonated Schiff base chromophore (see figure 7.4). The intrinsic properties of the native bacteriorhodopsin protein make it an outstanding candidate for use in optically coupled devices. These properties include: (1) long-term stability of the protein to thermal and photochemical degradation, (2) picosecond photochemical reaction times (both the forward and reverse photoreactions produce stable products in less than 5 psec), (3) high forward and reverse quantum yields permitting the use of low light levels for switching, (4) wavelength independent quantum yields, (5) a large shift in absorption spectrum accompanying photochemistry that permits accurate and reproducible assignment of state, (6) high two-photon cross sections permitting implementation of two-photon three-dimensional memories, (7) high second-order hyperpolarizabilities that open up nonlinear optical applications, and (8) the ability to form thin films of bacteriorhodopsin with excellent optical properties using the Langmuir-Blodgett or polymer matrix spin-coating techniques.

Plate 16 shows a model of the primary photochemical event in bacteriorhodopsin including voltage contours associated with the change in geometry. The geometry of the chromophore in bR is shown in green and the geometry of the photoproduct K is shown in gold. Green light drives the forward reaction (bR→ K) and red light drives the reverse reaction (K → bR). These two forms interconvert back and forth with a remarkable overall efficiency (80%) and speed ($\sim 5$ psec—the time it takes light to travel 1.5 mm or one-tenth of an inch). Note that only a small portion of the chromophore changes its geometry and that the chromophore is only a small portion ($\sim 10\%$) of the total protein system. The chromophore carries a positive charge that is primarily localized in the region that undergoes the geometry change. The primary event moves the charge about one Ångstrom (see inset in bottom middle of plate 16); the blue contours indicate negative voltage shifts, and the red contours indicate positive voltage shifts. The proton-pumping mechanism within the protein produces additional voltage transients that generate a separate and larger voltage signal. Some optical devices can take advantage of the voltage change by placing optically transparent electrodes across the thin films and by using an applied voltage to store or, upon reversal, erase an entire section of memory. The combined ability of the protein to respond to both light and applied voltage makes bacteriorhodopsin a unique electro-optical material.

### Cryogenic Digital Devices

The native protein, when isolated from the purple membrane and cooled to 77 K, displays the following photochemical equilibrium:

$$\text{bR}(\lambda_{max} = 576 \text{ nm}) \underset{\Phi_2 \sim 0.95}{\overset{\Phi_1 \sim 0.65}{\rightleftharpoons}} \text{K}(\lambda_{max} = 620 \text{ nm})$$

where bR represents the light-adapted form of the native protein, K represents the primary photoproduct, and $\Phi_1$ and $\Phi_2$ represent the forward and reverse photochemical quantum yields, respectively.

### Optical Random Access Memory

Optical RAM prototypes based on thin films of bacteriorhodopsin and the optical design shown in plate 17 have been devised with worst case access times of $\sim$ 40 nsec. Although this access time is slower than ultimately desired, the weak link in the current design is associated with latencies inherent in the optical XY scanner and not with the response latency of the protein. Picosecond experiments indicate that the design latency of a bacteriorhodopsin thin film at 77 K is less than 5 psec. The photoactive thin film contains light-adapted bacteriorhodopsin in a spin coated polymer film deposited on a mirrored substrate. Our research concentrates on enhancing the optical system so that we can decrease the worst-case access time to 1–4 nsec. The reliability of our design has been analyzed in detail (see figure 7.2).[5]

A schematic diagram of the principal components of a single memory plane of a bacteriorhodopsin-based optical random access memory is shown in plate 17. Two different designs are shown. The design shown in the upper diagram stores information in reflection mode on a photochromic condensing mirror (PC/CM). The design shown in the lower diagram stores information in reflection mode on a photochromic flat mirror (PC/FM) and requires an additional component (lens L2) to redirect the scanned reflected light onto the photodiodes (D1 and D2). Laser excitation to drive the photochemistry is provided by continuous wave red ($\lambda$ = 633 nm) and green ($\lambda$ = 543 nm) helium-neon lasers; the two beams travel down the vertical "CW laser bus." Prisms 1 and 2 (P1 and P2) are coated so as to couple $\sim$ 1 mW of the laser power into the optical shutters (OS1 and OS2). Thus, P1 and P2 require different reflective coatings depending upon the position of the memory plane assembly along the optical bus. The two beams are made collinear via P3 and P4. Lens L1, with a focal length roughly equal to the distance between L1 and the center of the photochromic surface (PC/CM or PC/FM), provides focused irradiation (spot diameter of $\sim$ 2 $\mu$m). The polymer film containing bacteriorhodopsin is deposited on either a condensing mirror (PC/CM) or a flat mirror (PC/FM), and the reflected laser light is directed onto the high-speed photodiodes (D1 and D2). A dichroic beam splitter (DS1) directs the reflected green laser beam to D1 and the reflected red laser beam to D2. Scanning of the focused laser beams across the photochromic surface is accomplished by using electro-optical scanners OPY and OPZ. Note that scanning in the plane of the page is shown to indicate the optical path as a function of this scanning operation. A memory word is represented by as many memory planes as there are bits in the word (a machine dependent number) plus two, which are used for reference bits. Thus, a word is read in parallel so that access time is determined by the single bit access latency, which is determined primarily by the speed of the electro-optical scanners.

## Ambient and Low-Temperature Analog Devices

The native protein, when isolated from the purple membrane and cooled to $\sim$ $-40$ C, or a modified protein generated by incorporating a chromophore analog, displays the following photochemical equilibrium:

$$bR(\lambda_{max} = 570 \text{ nm}) \underset{\Phi_2 \sim 0.95}{\overset{\Phi_1 \sim 0.65}{\rightleftharpoons}} M(\lambda_{max} = 410 \text{ nm})$$

where bR represents the light-adapted form of the modified protein, and M represents the blue-shifted (deprotonated) intermediate (see figure 7.2). The absorption maxima are approximate because these values are altered by environment and chromophore analog.

The effects of external electric fields on the physical and photophysical properties of bacteriorhodopsin have been an area of active study. An external field can induce both dichroic (anisotropic) and chemical (isotropic) changes in the absorption spectrum. An external field will also mediate the formation and decay of photocycle intermediates in a fashion that is proportional to the extent of charge translocation during the transition. The M → O transition involves the largest charge displacement and the application of an external field to an oriented sample can either increase or decrease the lifetime of M depending upon the field orientation. Under the proper environmental conditions, one can use electric fields to block the formation of the K photoproduct or to decrease the rate of decay of the M intermediate, without slowing down its rate of formation. The ability to adjust the lifetime of M by using an external field allows for voltage control of the transmittance of bR thin films. This flexibility is important in the design of spatial light modulators and holographic associative memories.

## Spatial Light Modulators

Recent research in optical engineering has demonstrated the unique capability of two-dimensional optical processing systems to perform complex mathematical and processing functions such as pattern recognition, image processing, solution of partial differential and integral equations, and matrix-vector and matrix-matrix linear algebra and non-linear arithmetic. Interest in exploring the complex architectures associated with optical processing is due to the inherent speed and unique functionalities that derive from the parallel-processing and interconnection capabilities of optical systems. Spatial light modulators (SLMs) are integral components in a majority of one- and two-dimensional optical processing environments. These devices modify the amplitude, intensity, phase, and/or polarization of a spatial light distribution as a function of electrical signals and/or the intensity of a secondary light distribution.

The observation that a thin film of bacteriorhodopsin can act as a photochromic and photorefractive bistable optical device (either bR ⇔ K or bR ⇔ M photoreactions) or as a voltage-controlled bistable optical device (bR ⇔ M photoreaction) suggests that it has significant potential as the active medium in SLMs. Soviet scientists have carried out a number of investigations that support this enthusiasm. The absorption spectra of the key intermediates of the bR photocycle are shown in figure 7.4. Under the appropriate environmental conditions, or with the appropriate chromophore analog substitutions, the lifetime of the M intermediate is dramatically increased (see previous discussion). The bR ⇔ M photoreaction then provides the necessary attributes for a highly flexible photochromic spatial light modulator. For example, an adjustable two-dimensional threshold modulator can be prepared that uses a thin film of bR enclosed between two transparent electrodes. If an image is imposed on the device using monochromatic light from a helium-neon laser (632.8 nm), the light is absorbed by bR ($\epsilon_{bR}$(633 nm) @ 20,000 $M^{-1}$ $cm^{-1}$) but not by M ($\epsilon_M$(633 nm) < 500 $M^{-1}$ $cm^{-1}$). The spatial intensity distribution of light exiting the thin film will be altered by the intensity pattern so that high-intensity segments are enhanced relative to lower intensity segments due to photochromic processes. Thus, the thin film acts as a threshold device. The threshold level can be adjusted by changing the voltage across the two transparent electrodes, which alters the lifetime of M (figure 7.5).

**Holographic Associative Memories**
Bacteriorhodopsin exhibits all of the characteristics required of an eras-Bacteriorhodopsin exhibits all of the characteristics required of an erasable holographic memory. When compared to other photochromic films currently in use, relative performance is on average superior. There are two key attributes of bR that yield advantages over the competitive photochromics. First, the cyclicity (the number of times the compound can be photochemically cycled without degradation in performance) is significantly higher for bR than most other photochromics. The origin of this characteristic is due in part to the protective attributes of the protein-binding site, an environment that has been optimized through evolutionary processes to enhance chromophore stability. Second, the variability of the storage time provides for design flexibility. While the holographic efficiency (3–4%) is lower than might be desired, it is ample for most applications.

One application under development is the use of thin films of bR as the photoactive components in Fourier transform holographic (FTH) associative memories. Our current design is shown in figure 7.6 and is based on the closed loop autoassociative design of Paek and Psaltis.[6]

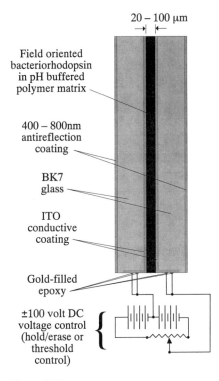

20 – 100 μm

Field oriented
bacteriorhodopsin
in pH buffered
polymer matrix

400 – 800nm
antireflection
coating

BK7
glass

ITO
conductive
coating

Gold-filled
epoxy

±100 volt DC
voltage control
(hold/erase or
threshold
control)

**Figure 7.5**
Schematic diagram of a voltage controlled spatial light modulator based on a
polymer film containing oriented light-adapted bacteriorhodopsin. The application of a voltage across the thin film increases (or decreases) the lifetime of
the M intermediate by opposing (or enhancing) the proton translocation process. This device can be used as the threshold device in the holographic associative memory described in the text.

The basic elements of the optical design are described below. FTH
associative memories have significant potential for applications in optical
computer architectures, optically coupled neural network computers,
robotic vision hardware, and generic pattern recognition systems. The
ability to rapidly change the holographic reference patterns via a single
optical input—while maintaining both feedback and thresholding—increases the utility of the associative memory and, in conjunction with
solid state hardware, opens up new possibilities for high-speed pattern
recognition architectures. If the bR-based holographic memory plane is
prepared using photographic gelatin mixed with polyvinyl alcohol and

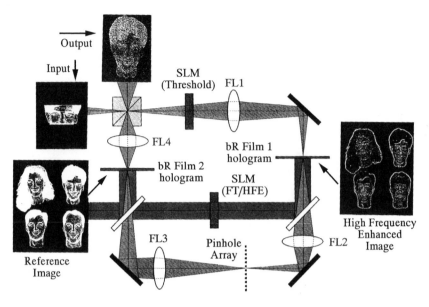

**Figure 7.6**
Schematic diagram of a Fourier transform holographic (FTH) associative
memory with read/write FTH reference planes using thin polymer films of bacteriorhodopsin to temporarily store the holograms.

the photochromic material is enclosed between two transparent conducting electrodes, a square wave voltage pulse can be used to hold the
hologram and a reversed pulse can be used to enhance the erase cycle.
An alternative approach is to use a light pulse to erase the hologram.
The diffraction-limited performance of the bR films, coupled with high
write/erase speeds associated with the excellent quantum efficiencies of
the polymer films, represents a key element in the potential of this
associative memory device.

The Fourier transform holographic (FTH) associative memory operates as follows. The input image enters the system through a beam
splitter at the upper left of the figure and illuminates a spatial light
modulator (SLM) operating in threshold mode. The light reaching the
SLM is the superposition of all the images that have been stored in the
multiplexed holograms. Each image is weighted by the inner product
between the recorded pattern from the previous iteration and itself. The
pinhole array is designed so that the pinholes (diameter = 500 mm) are
aligned precisely with the optical axes of the multiple FTH reference

images stored on the two bR films. The FTH reference images enter from the middle left, and a separate Fourier transform high-frequency enhancement (FT/HFE) SLM is used to edge enhance the images on bR film 1 in order to enhance the autocorrelation peak. The pinholes are used to eliminate ghost holography, but in the process the spatial invariance of the input image reconstruction is lost. Thus, proper registration of the input image is required for proper associative output. The output image is a full reconstruction of the image stored on the FT hologram that has the highest correlation with the input image (i.e., produces the largest autocorrelation flux through its aligned pinhole). Thus, only a partial input image is required to generate a complete output image. The thresholding SLM is critical to this application as its response level must be dynamically adjustable in order to compensate for variable autocorrelation fluxes.

## Notes

I would like to thank the graduate students (Jim Bennett, Tom Cooper, Anne Marie Farone, Rick Gross, Susan Hom, Lynn Hubbard, Elaine Hyde, John Izgi, Lynn Hubbard, Mark Masthay, Lionel Murray, Leonore Findsen, Jeff Stuart, Brian Pierce, Jack Tallent, Chian-Fan Zhang), postdoctoral fellows (Drs. Paul Fleitz, Alan Schick, Cora Einterz), and collaborators (Drs. David Bocian, James Dabrowiak, Janos Fendler, Rick Lawrence, Jonathan Lindsey, Laurence Nafie, Koji Nakanishi, Charles Spangler, Benjamin Ware) who contributed to the research reported here. This research was supported in part by grants from the W.M. Keck Foundation, the National Science Foundation, the National Institutes of Health, the Office of Naval Research, Rome Laboratories (USAF), and Digital Equipment Corporation.

1. See G. Schick, A. Lawrence, R. Birge, "Biotechnology and molecular computing," *Trends in Biotechnology*, 6 (1988): 159–163. "High electron mobility technology: The technology for decreasing material electrical resistance (increasing electron mobility) in solid-state devices by chemically modifying the solid-state lattice structure in order to reduce electron scattering mechanisms."

2. "Josephson device: A device that operates on the basis of the Josephson effect, i.e the spontaneous tunneling of current through an insulating barrier that separates two super-conducting materials." Ibid.

3. M. La Brecque, "Molecular electronics: circuits and devices a molecule wide," and "Molecular electronics: devices that assemble themselves," *MOSAIC* (The National Science Foundation) 20 (Spring 1989): 1. These articles present a brief, general overview of molecular electronics.

4. R. Birge, A. Lawrence, "Quantum effects, thermal statistics and reliability of nanoscale molecular and semiconductor devices," *Nanotechnology*, 2 (1991): 73–87.

5. R. Birge, "Photophysics and molecular electronic applications of the rhodopsins," *Annu. Rev. Phys. Chem.* 41 (1990): 683–733. This article discusses bacteriorhodopsin and its use in optical memory devices.

6. E. Paek, D. Psaltis, "Optical associative memory using Fourier transform holograms," *Opt. Eng.* 26 (1987): 428–433.

## Discussion

*Drexler:* We know from the example of DNA that it is possible with a more or less mechanical system to write, store, and read two bits per base (i.e., per monomeric molecule) stably over a period of time. Are there basic physical reasons why you see worse performance in molecular electronics?

*Birge:* The chemical processes by which DNA is read are dynamic processes with time for error correction. Furthermore, the act of reading DNA is nondestructive, and hence the chemical processes can operate in an integrating mode which enhances reliability. Our optically coupled devices require ensemble averaging because the act of reading disturbs the state of the system. I would not agree that ensemble averaging should be characterized as leading to "worse performance," but I will agree that further work on fault-tolerant molecular-based systems is warranted and would yield significant potential.

*Audience:* You showed a relationship between the size of the ensemble and the error rate at two different temperatures. Was there a difference between the two temperatures?

*Birge:* If one lowers the temperature of the system, one decreases thermal noise and kinetic energy which can change the molecular state or break bonds which are critical to performance. In general, therefore, the lower the temperature the higher the reliability. Each system is unique however, and there are no simple rules that can be used to predict a priori the quantitative advantages of low-temperature operation. The optical memories based on bacteriorhodopsin are particularly complicated in this regard because changing the temperature changes the molecular states that represent the binary states of the ensemble.

# 8

# Quantum Transistors and Integrated Circuits

Federico Capasso

Bandgap engineering, molecular beam epitaxy, and nanofabrication techniques have made possible the invention and realization of a new class of quantum devices. This chapter focuses on quantum devices based on resonant tunneling through quantum wells. These structures hold promise for greatly reducing the complexity of circuits in several digital and analog applications.

## Groups III and V Semiconductors

Silicon, a group IV element, dominates the semiconductor industry and will always be a basic engineering material, like steel and plastic. The semiconductors discussed in this paper are not silicon but are instead the more exotic materials from groups III and V of the periodic table.

The most important property of a semiconductor is the energy band gap, which is the energy difference between the valence band and the conduction band of a crystalline solid.[1] You can vary the conductivity of a semiconductor over orders of magnitude by doping or adding impurities. If you add impurities that have a valence higher than that of the host atoms (doping with electrons), or impurities that have a lower valence (doping with positive charges, or "holes"), you can change the conductivity over many orders of magnitude. Figure 8.1 illustrates the relationship between the energy band gap and lattice constant (the size of the crystalline unit cell) of several semiconductors.

This chapter focuses on the combination of "lattice-matched alloys" defined by vertical lines in figure 8.1. Lattice-matched alloys are alloys constructed from two or more semiconductors with the same lattice

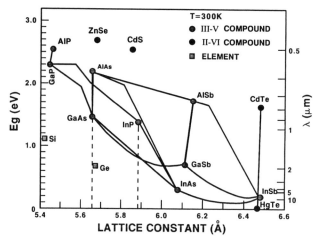

**Figure 8.1**
Energy gap (eV) vs lattice constant (Å) of various semiconductors.

constant. Thus, on top of a gallium arsenide (GaAs) substrate, you can grow alloys of aluminum-gallium-arsenide (AlGaAs) with the same lattice constant as the GaAs substrate. These semiconductor combinations will be very important for the electronics of the future (possibly the near future). Another lattice-matched alloy is indium-gallium-arsenide-phosphide (InGaAsP), which can be grown on indium phosphide (InP). This alloy is important for light-wave communication devices but will not be discussed in this chapter.

These lattice-matched alloys are the playground of the solid-state physicist and device physicist.[2] If you have the right tools, you can use these materials to create devices that exhibit quantum mechanical behavior. This field was characterized by Leo Esaki as "do-it-yourself quantum mechanics." Esaki shared the Nobel Prize in physics in 1973 for his pioneering work on tunneling in solids.

**Molecular Beam Epitaxy**

Molecular beam epitaxy (MBE) is a technique used to grow materials with atomic scale control, including the doping composition. MBE can be used to engineer new device building blocks for new electronic and

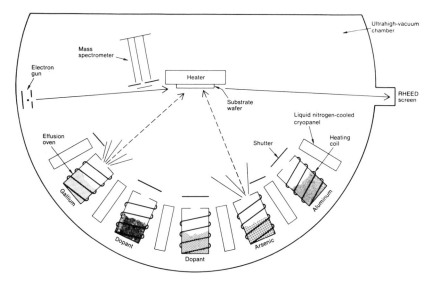

**Figure 8.2**
Diagram of MBE machine.

optical applications. Drs. Alfred Cho and John Arthur of the solid-state electronic research group at Bell Labs pioneered this work in the late sixties.

The basic framework for an MBE apparatus is a high-vacuum chamber (figure 8.2). To grow the alloy, gallium-aluminum-arsenide (GaAlAs), the three elements are placed in the cells as indicated and heated at very high temperatures so that an effusion process forms molecular beams of the elements. Other cells are used to provide dopants. A computer-controlled shutter turns the beam on or off to deposit precisely controlled layers. The substrate wafer is heated (typically in the range of 400° to 700° C, depending on the material) so that molecules reaching the substrate surface move freely and eventually replicate the crystalline structure of the substrate surface. In a lattice-matched alloy you can thereby maintain the lattice of the substrate throughout the structure.

Various analytical tools such as a mass spectrometer can be added to the basic setup, and an entire MBE system costs between $900,000 and $1.5 million. MBE is used at Bell Labs and at other companies for manufacturing GaAs integrated circuits.

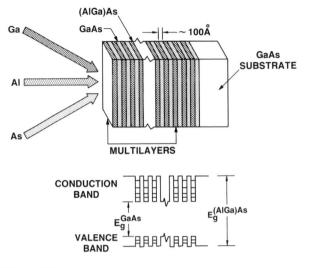

**Figure 8.3**
Energy band diagram for multiquantum well structures of AlGaAs and GaAs.

## Types of Quantum Devices

MBE can be used to produce multiquantum well structures by using the molecular beams to form periodic alternating layers (each typically 10 to 100 Å thick) of two different materials.[3] The physics of these alternating layers can be described in a very simple way using energy band diagrams. Figure 8.3 displays such a structure and its energy band diagram. It consists of alternating layers of aluminum-gallium-arsenide (AlGaAs) and gallium arsenide (GaAs). Energy levels are shown for each layer for the conduction band and for the valence band.

The crucial point is that if you have a layer thickness comparable to the de Broglie wavelength, then the electron motion is quantized perpendicular to the layers of the superlattice to form energy levels in the structure. (For GaAs at room temperature, the thermal de Broglie wavelength is approximately 250 Å). The separation between energy levels varies as one over L squared, where L is the thickness of each layer. Using this technique you can engineer many materials with variously shaped potential wells of the kind studied in quantum mechanics. That is what is meant by "do-it-yourself quantum mechanics."

p-n JUNCTION

**Figure 8.4**
Simple MBE device, p-n junction.

Electrons are still free to move parallel to the layers, so you really only have energy levels—or what look like energy levels—perpendicular to the layers. Parallel to the layers there are energy bands. The energy levels observed perpendicular to the layers are really the bottoms of quantized energy bands described by band structure theory. However, for most practical purposes, you can use one-dimensional band diagrams and consider the energy levels as you would consider energy levels in atoms.

A p-n junction, a simple device that can be made by MBE, is shown in figure 8.4. This conventional device, one that has been known for 40 years, is made by doping a semiconductor to give an n-type and a p-type layer, creating a barrier that can be used to make a transistor.

More exciting, however, are the quantum structures shown in figure 8.5. The first is a quantum well, which is a potential energy well (as previously discussed). The second is a double barrier well, separated from the environment by thin barriers so that you can tunnel into and through the quantum states confined therein. Third is a structure that uses coherent transport over many layers, a superlattice. The superlattice, pioneered by Leo Esaki and Ray Tsu at IBM in the late 1960s,[4] is created

QUANTUM WELL        DOUBLE-BARRIER        SUPERLATTICE

**Figure 8.5**
Energy diagrams of simple quantum structures.

by coupling many quantum wells. By overlapping the energy states of several quantum wells, a superlattice eliminates the degeneracy of the energy levels of their individual wells and forms the superlattice energy bands shown. These three quantum structures provide new building blocks that we can use to create new device functions.

## Fabrication of Quantum Devices

Plate 18 shows a high-resolution transmission electron microscope (TEM) image of a gallium-indium-arsenide/indium-phosphide (GaInAs/InP) quantum well. The well consists of three molecular layers of GaInAs sandwiched between two InP layers. Each red-yellow dot represents a tunnel between atoms. The hetero-interfaces, the boundaries between the layers, are atomically sharp. With MBE you can grow the precise atomically defined interface required to get quantized motion in the vertical direction. An uneven interface would create scattering that would destroy the coherent interference effects needed to obtain the desired energy levels.

The quantum well shown in figure 8.5 does not show the periodicity of a true lattice because it consists of only one layer sandwiched between two layers of a second material. One can, however, make a superlattice by alternating layers of the two materials (InP and GaInAs).

By varying the period and the composition of the superlattice you can engineer new properties not present in the original material. (For example, MBE can be used to engineer asymmetric potential wells to create structures with extremely large value for $\chi_2$. See the final section of chapter 4 by Michael Ward, "Design of Self-Assembling Molecular Systems," for a discussion of the optical properties $\chi_2$ and $\chi_3$. This chapter focuses on electronic devices, but the range of devices you can create with MBE is wide open.)

## Resonant Tunneling Diodes

Resonant tunneling is an idea nearly as old as quantum mechanics itself. It was first proposed as an explanation for alpha ray radioactivity or induced nuclear reactions by Gamow in the Soviet Union and Condon in the United States in 1928.

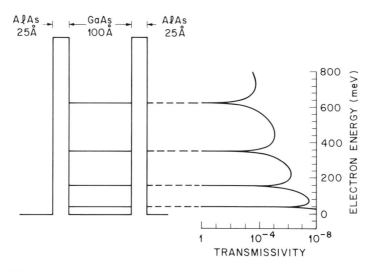

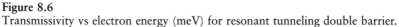

**Figure 8.6**
Transmissivity vs electron energy (meV) for resonant tunneling double barrier.

You can use resonant tunneling structures to create several interesting applications. A quantum well can be formed with very thin barriers to create the energy levels as shown in figure 8.6. This effect is equivalent to the interference phenomenon seen when two semitransparent mirrors are facing each other. If you try to transmit light of different colors, the transmission of the mirror will peak at certain wavelengths such that a semi-integer number of wavelengths is equal to the distance between the two mirrors. When the distance between the two barriers of a quantum well equals a semi-integer number of de Broglie wavelengths, energy states are created. If you measure the transmission coefficient as a function of incident kinetic energy perpendicular to the barrier, there are peaks in the transmission that correspond to the energy levels of the well because of constructive interference of the transmitted waves at certain energies. (There are complications at high-current densities due to the Pauli Exclusion principle that we will ignore here.)

It is difficult in practice to vary the energy of incident carriers in a resonant structure. You need to approach the problem from a different angle. Figure 8.7 shows the energy band diagram of a resonant tunneling diode and a schematic of its current-voltage characteristic. Since it is

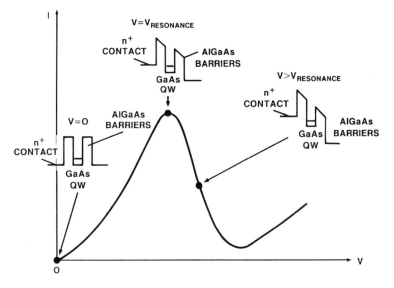

**Figure 8.7**
Energy band diagrams and schematic current-voltage characteristics for a resonant tunneling diode.

difficult to vary the incident energy, you can apply a voltage to the structure to vary the energy of the eigenstates in the well instead. In this way you can move the energy level up and down in comparison to the incident energy. At some critical voltage such that the energy level is lined up with the bottom of the Fermi sea of carriers in the emitter layer, you get a peak in current due to resonant tunneling. If you lower these energy states below this critical point, the current decreases due to conservation of lateral momentum and energy during tunneling. In theory, the current goes to zero, but in reality, there are always inelastic processes that prevent the current from vanishing. The pioneering studies in this area were by Esaki's group at IBM in the early 1970s.

The current experimentally obtained from such a device at 300 K is shown in figure 8.8. In this example the structure consists of a layer of aluminum-indium-arsenide (50 Å), and a layer of a second alloy, gallium-indium-arsenide (50 Å), followed by another layer of the first alloy (50 Å), all grown by MBE. The device displays a negative differential of resistance at some current; that is, at some voltage the current peaks

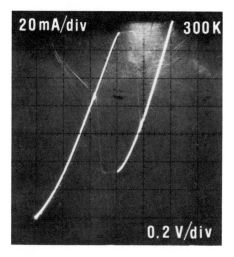

**Figure 8.8**
Current (i) vs voltage (V) for resonant tunneling diode (experimental data).

and then decreases. Thus you can use this device to make an oscillator.

However, logic circuits require a transistor, a device with three terminals, instead of a simple diode with only two terminals. Thus, the question was how to make a resonant tunneling transistor.

Recently I have been studying some new devices, including a resonant tunneling transistor. We and others can now reproduce the current peak within one or two kT,[5] which is adequate for many circuit applications, and much better than in a tunneling diode.

The question that has guided our attempts to exploit quantum well structures has been, "Can we construct a logic circuit such that a given energy state of a quantum well corresponds to a logic level of our transistor?" Although these quantum well structures probably will not replace binary circuits in the near future, the question remains, "What kind of architectures can be built from quantum transistors?"

To achieve resonant tunneling through many energy levels, we have used MBE to engineer devices with a wide range of quantum potential wells. We have synthesized an artificial potential well that acts like a harmonic oscillator (figure 8.9a). The harmonic oscillator is a universal type of potential in quantum mechanics—it appears in the electromagnetic field and it appears in molecules. My group at Bell Labs has

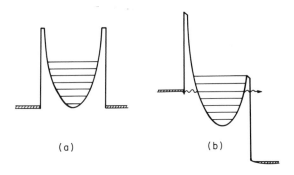

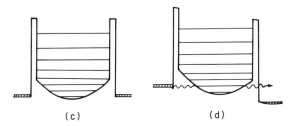

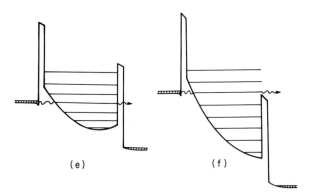

**Figure 8.9**
Energy band diagrams of parabolic potential well.

synthesized such a parabolic potential well with equally spaced energy levels. You can probe the energy states with resonant tunneling by applying a bias voltage (figure 8.9b). Further, quantum mechanics allows quantized states (resonances) also in the classical continuum above the barriers (figure 8.9f). This is because in quantum mechanics it is possible to get a reflection above the barrier (this is unlike classical mechanics). The barrier material in the wells is aluminum-arsenide (AlAs) or aluminum-gallium-arsenide (AlGaAs), and the well itself consists of compositionally graded aluminum-gallium-arsenide AlGaAs.

Working with Art Gossard (now at the University of California, Santa Barbara) we built a parabolic quantum well with a width of 400 Å. Figure 8.10 displays the voltage and current data from this device and shows up to 15 quantum mechanical resonances. This demonstrates the feasibility of engineering resonant tunneling diodes with a variety of quantum states.

### Quantum Transistors Using Resonant Tunneling

A diode is not a transistor, and you need transistors to make truly useful devices. How can you incorporate resonant tunneling into a transistor? Figure 8.11 displays the energy band diagram for Shockley's original bipolar transistor. The energy is shown along the distance coordinate for the sandwich arrangement of an emitter, a base, and a collector. The emitter and collector are made of a material with n-type doping. They contain impurities that produce an excess of electrons. In the middle is the base, with p-type doping. It contains impurities that produce a lack of electrons, or holes. This produces an n-p-n junction or a "bipolar transistor."

Bipolar transistors work when a voltage bias is applied between the base and the emitter, thus decreasing the energy barrier, whereas an opposite bias between the base and the emitter increases that energy difference. If a signal is sent across the junction between the emitter and the base, a small variation in the height of the energy barrier between emitter and base causes a large variation in the amount of current that flows between emitter and collector. This is due to the Boltzmann distribution: a small decrease in the barrier produces a large increase in the number of charges with the necessary thermal energy to be carried across

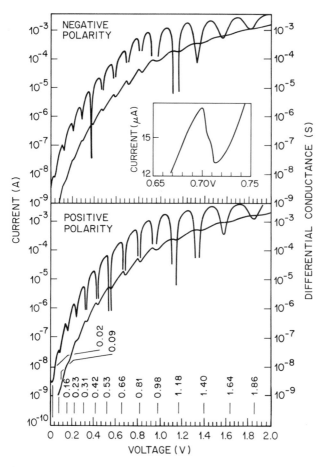

**Figure 8.10**
Voltage (V) vs current (A) of parabolic well resonant tunneling diodes.

the barrier.[6] Thus the transistor is an amplifying device, since a small input signal gives a large variation in the output.

In 1985, I proposed a transistor that would incorporate resonant tunneling.[7] Figure 8.12 shows the energy band diagram of such a transistor. By adding quantum sandwiches to the base of the transistor, energy states lie in the path of electrons going from the emitter to the collector. These double barriers act as energy filters that only let electrons pass for certain bias conditions. Figure 8.12a shows the quantum transistor at zero bias. As bias voltage is applied (figure 8.12b), current flows

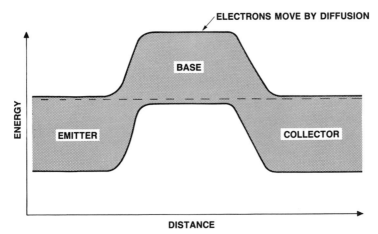

**Figure 8.11**
Energy band diagram of conventional bipolar transistor.

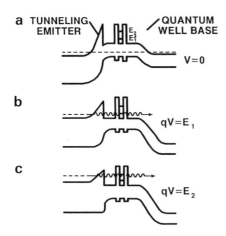

**Figure 8.12**
Resonant tunneling transistor at three voltages.

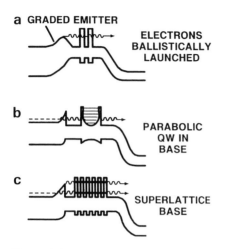

**Figure 8.13**
Alternative designs for resonant tunneling transistors.

(as with a conventional transistor), reaching a peak value when the voltage is increased to achieve resonance with the first energy state ($E_1$). As voltage is further increased, the current drops, then peaks again at each successively higher energy state (figure 8.12c).

Transistors can be engineered with a multitude of different quantum energy well profiles built into the bases, as shown in figure 8.13. These devices can be used to study quantum transport through artificially grown materials, such as superlattices. The base of these devices is typically a few thousand Ångstroms thick.

In theory, increasing the bias voltage until it reaches resonance with the second, third, fourth, and so on, energy levels should produce multiple peaks. These peaks could then be associated with various logical states. In fact, this was very difficult to do.

A few years ago we announced the first multistate transistor.[8] This device, shown in figure 8.14, employs multiple double barriers in the emitter rather than in the base. An increase in the base emitter voltage leads to sequential suppression of resonant tunneling through the different double barriers, giving rise to multiple peaks in the current. This is a true multistate device because the logic states correspond to different physical states of the structure.

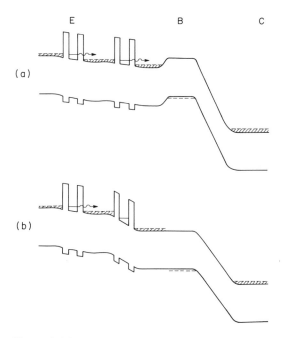

**Figure 8.14**
Energy diagram of multistate transistor with two double barriers in the emitter.

Figure 8.15 displays the experimental data obtained from the transistor shown in figure 8.14 operating at 77 K and 300 K. There are two peaks in the output current as a function of the input voltage for this device due to the double barriers in the emitter. These two output peaks represent the two "on" states. The transistor is "off" at low output currents. The gain of this transistor is 60 at room temperature.[9] This transistor has been operated at 25 GHz.

## Applications of Multistate Quantum Transistors

What can you do with a multistate transistor? The reproducibility of the two peaks in figure 8.15 is good enough to try making circuits. There are a few classes of applications for which these quantum devices might have a real edge in the near future. The key advantage of resonant tunneling transistors is that they allow you to replace many conventional

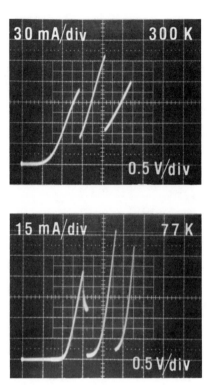

**Figure 8.15**
Current-voltage characteristic for multistate transistor with two double barriers in the emitter operating at 300 K (top) and 77 K (bottom).

transistors (up to several tens) in many applications. Some of the interesting applications of quantum transistors include the following:

• Analog circuits, such as frequency multipliers, analog-to-digital converters, and oscillators
• Logic circuits, such as parity generators
• Multiple-state memories

**Analog Circuits**

You can use a device such as a quantum transistor to create a simple frequency multiplier because it has two peaks in its transfer function (input versus output). A schematic for a frequency multiplier that produces an output current multiplied in frequency by a factor of three (for

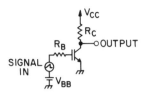

**Figure 8.16**
Schematic of frequency multiplier.

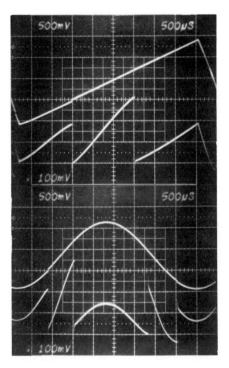

**Figure 8.17**
Input and output signals for the frequency multiplier shown in figure 8.16, in the case of triangular wave input (top) and sine wave input (bottom).

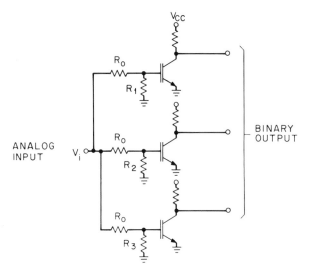

**Figure 8.18**
Schematic for analog-to-digital converter.

a triangular input signal) or five (for a sine wave input signal) is shown in figure 8.16. The experimentally measured output signals for this device are shown in figure 8.17 for triangular and sine wave inputs. This device has been operated at frequencies of up to a few GHz. This is clearly an area where quantum devices will make an impact.

Multistate transistors can also be used to create analog-to-digital converters. Figure 8.18 shows a schematic of a simple A-to-D converter that uses three multistate quantum transistors in parallel. The analog input signal varies in time, creating a waveform. Because the resistors in the voltage scaling network ($R_1$, $R_2$, $R_3$) have different values, each transistor (connected to one of the scaled resistors) receives a different voltage input. Thus some transistors are on while others are off. This produces a binary output from the analog input. We should be able to build A-to-D converters based on this design that will function at up to 30 GHz. These devices could have a great impact in a particular niche of circuit technology. However, in contrast to these specialized applications, it is not now realistic to think of building very complex integrated circuits with quantum devices.

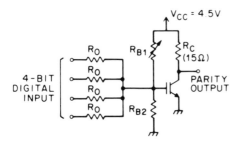

**Figure 8.19**
Schematic for parity bit generator.

## Logic Circuits

Bell Labs is very interested in communication systems, and these systems use a fascinating circuit called a parity bit generator that tells how many errors have been made in transmitting digital data. Since no transmission can be totally error free, you need a circuit that tells, for each digital word, whether or not an error has been made during transmission. A parity bit generator measures the signal and tells you whether you have an odd or even number of (binary) "ones" in the word. To build a parity bit generator for a four-bit digital word, an analysis by Boolean algebra shows that you need a minimum of 24 conventional transistors.

Using an architecture based on multistate quantum transistors (figure 8.19), we have built a device that uses only one transistor rather than 24. The input to the device is four bits. A single one, or three ones, at the input corresponds to a high current passing through the transistor, giving a low voltage output. The multistate transistor thus gives a low voltage when there is an odd number of ones at the input and a high voltage when there is an even number of ones at the input. Figure 8.20 shows the measured data from operating this device at room temperature.

This type of savings may not be possible with all circuits. We are proceeding empirically because we do not yet have general theorems that allow us to expand a general circuit function in terms of simple functions. We are looking for help from people in computer science to develop such theorems.

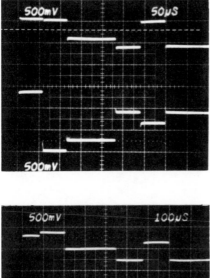

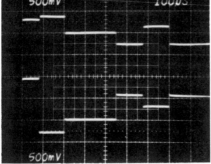

**Figure 8.20**
Output signals for parity bit generator.

## Multistate Memories

Computer scientists in the past have been reluctant to consider the possibility of multistate devices because they were not available. A basic theoretical question is what are the logic gates in a multivalued logic system that would be equivalent to NAND and NOR gates such that any particular logic function could be described in terms of these gates. This issue of completeness remains an unresolved issue.

Another exciting application is multistate memories. It should be possible to use quantum transistors to make a memory with multiple stable states by utilizing the two peaks in the transistor's transfer function.

## The Future of Quantum Devices

You can build novel quantum semiconductors that can perform certain functions in a much simpler way or that can perform new functions. But the question arises, "Can you not only quantize the motions of electrons in the direction of epitaxial growth but also in the plane?" Achieving quantization in additional dimensions is a challenge of crucial importance for the future of quantum devices.

All transistors discussed in this chapter are bipolar transistors. But transistors manufactured in volume are not bipolar. Integrated circuits use MOSFET devices, and MOSFET architecture is surface based.[10] A MOSFET transistor has three contacts on a surface, and the current parallel to the surface is controlled by a gate. The scale of the lithography determines the minimum feature length of the transistor or the length that the current is transported through the transistor. Optical lithography can produce features as small as 3000 Å (0.3 microns) using lasers in the ultraviolet region of the spectrum. Silicon-based electronics is expected to reach its limits somewhere between 1000 and 2500 Å. How will the electronics industry continue to evolve if it cannot scale devices smaller than this limit? Perhaps by exploiting quantum effects, but I believe we have to be very cautious about predicting how this will happen.

However, even if there is a question about our ability to exploit quantum microelectronics, there is some beautiful physics that can now be done with quantum structures that exploit motion in the x–y plane, not only in the (vertical) direction of epitaxial growth. MBE provides a technique for creating quantum wells that quantize electron motion in the vertical direction. With the help of lithography, you can create structures that quantize motion in two dimensions so that there is free motion in only one direction. This is the equivalent of a single-mode optical fiber, a "quantum wire." A further extension would be to make a "superatom" or "quantum dot." This would be a microstructure in which the electron motion is quantized in three directions to make an "atom" of a few hundred Å dimension.

Henry Smith at MIT has done some beautiful work in X-ray nano-lithography creating 30 nm "fences" of polymethyl methacrylate

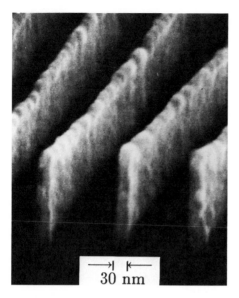

**Figure 8.21**
X-ray nanolithography of PMMA. (Courtesy H. Smith, MIT.)

(PMMA), a polymer (figure 8.21).[11] He has also made quantum wires and something very similar to a superlattice (the dimensions are not quite there yet) by forming a metal grid on the surface that potentially gives you quantization in the plane (figure 8.22). This work is leading to exciting new physics. Let us not forget that, for example, the quantum Hall effect discovery was made in a semiconductor materials device.[12]

## Conclusion

This chapter is intended to give you a flavor of this rapidly developing field. In summing up, consider a critical question, "How well can we control the current peaks in the transfer functions for these devices?" I am optimistic that we can control these very well for the sorts of circuits we are envisioning. We do not, however, claim to be able to make in the foreseeable future complex integrated circuits of a million devices on a chip because III–V materials cannot be manufactured in quality comparable to that achieved in silicon. However, quantum devices can have an impact in a niche of electronics that requires high performance.

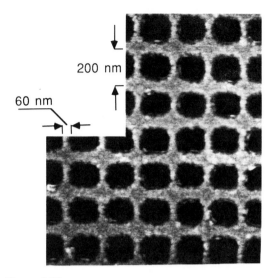

**Figure 8.22**
Ti/Au Schottky grid-gate[13] on GaAlAs/GaAs. (Courtesy H. Smith, MIT.)

Our demonstration projects have been very encouraging. We are currently working with a development group at Bell Labs to make some of these quantum transistors.

Band structure engineering is a beautiful playground for the solid state physicist. One has to be cautious, however, about predicting the time scale for these developments. The following quote from Richard Feynman, who was a visionary as well as a great scientist, serves as a reminder to those of us who work on the boundary between physics and technology: "For a successful technology, reality must take precedence over public relations for nature cannot be fooled."

## Notes

1. When atoms assemble into a crystalline solid, the discrete energy levels of an isolated atom are replaced by bands of allowed energy that the electrons in the solid can occupy, separated by gaps of forbidden energy. The energy band that the valence electrons occupy is the valence band, and the higher energy band in which electrons are free to diffuse throughout the solid is the conduction band. In an insulator, the valence band is completely filled and there are no electrons in the conduction band. Conductors have a partially filled conduction band. In semiconductors, the valence and conduction bands can be bridged by thermally excited electrons.

2. "Band-gap engineering" is a term used to describe this approach to the design of new semiconductor materials and devices. MBE and related techniques are used to create devices with band diagrams that have nearly arbitrary and continuous band-gap and doping variations. For example, you can use MBE to vary the composition of a semiconductor in the direction of epitaxial growth to create a gradient in energy band structure with distance.

See F. Capasso, "Band-gap engineering: from physics and materials to new semiconductor devices," *Science* 235 (9 January 1987): 172–176. For further reading on quantum devices, see *Physics of Quantum Electron Devices*, vol. 28, ed. F. Capasso (New York: Springer-Verlag, 1990).

3. A superlattice consists of several alternating layers of lattice-matched materials that have have different energy band structures. MBE can be used to create a superlattice by depositing a layer of one material on top of a layer of another material. A quantum well is formed if a thin layer (several hundred Å or less) of a material of low band gap is sandwiched between (barrier) layers of a material of higher energy gap. A superlattice is formed if several quantum wells are grown on top of one another with the barrier layers between the layers of low band gap material thin enough to allow tunneling to couple the wells. The alternating layers in a superlattice could be several to several tens of atomic layers thick, or, in the limiting case, only one atomic layer thick.

4. L. Esaki, R. Tsu, "Superlattice and negative differential conductivity in semiconductors," *IBM J. Res. Dev.* 14 (1970): 61.

5. T is the absolute temperature in degrees Kelvin (K), and k is the Boltzmann constant, or $1.38 \times 10^{-23}$ joule per K. The average kinetic energy of motion for an object is $(1/2)kT$ per degree of freedom.

6. The Boltzmann distribution applied here states that the increase in probability of finding an electron at a lower energy varies as the exponential of the decrease in energy divided by $kT$.

7. F. Capasso, R. Kiehl, "Resonant tunneling transistor with quantum well base and high-energy injection: A new negative differential resistance device, *J. Appl. Physics.* 58 (1985): 1366.

8. F. Capasso, S. Sen, A. Y. Cho, D. L. Siuco, "Multiple negative transconductance and differential conductance in a bipolar transistor by sequential quenching of resonant tunneling," *Appl. Phys. Lett.* 53 (1988): 106.

9. The gain of a transistor is defined as the ratio of the collector current to the base current.

10. MOSFET is the acronym for metal oxide semiconductor field effect transistor, the type used in integrated circuits.

11. See H. Smith, H. Craighead, "Nanofabrication," *Physics Today* 43 (February 1990): 24–30.

12. See K. von Klitzing, "The quantized hall effect," *Rev. Mod. Phys.* 58 (1986): 519–531 for von Klitzing's Nobel Prize lecture on the quantum Hall effect. See also B. Halperin, "The quantized Hall effect," *Scientific American* 259 (April 1986). Briefly, the quantized Hall effect is the surprising discovery made during the 1980s that certain resistance properties of semiconducting materials, measured at very low temperatures and at very high magnetic fields, gave extremely precise information about the charge carriers in the material. Under appropriate conditions a series of plateaus in the Hall resistance could be measured that precisely equaled the electron charge divided by Planck's constant times some integer or some simple fraction.

13. The term "Schottky grid-gate" is coined from the term "Schottky barrier," which refers to the potential barrier created at the interface between a metal and a semiconductor.

## Discussion

*Drexler:* I noticed that most of your conservatism and caution, which I believe are very well placed, are attached to questions of dates, which are always speculative, and to fabrication capabilities in pushing the limits of lithography and bulk processing. I would be very interested in whatever you can say that is not speculative with respect to physics, and not speculative with respect to materials properties, if you assume for the moment that you don't have any fabrication constraints. For example, assume that you can put whatever material you want wherever you want it, with no flaws, you can make the interfaces epitaxial if the lattice spacings and chemical compatibilities are suitable, and that you have control of the lateral dimensions as well as the growth dimension. Can you then say anything about sizes, speeds, energy dissipation, and other issues of concern to computer engineers?

*Capasso:* Your question assumes that you will eventually get all the engineering correct and then only be limited by the laws of physics. However, it has often been the case in the history of the development of technology that in struggling to develop your technology, you are forced to include new physical assumptions, thus limiting your predictive power. The physics will instead depend upon the technology and upon the detailed path of technological development.

With devices on the atomic scale, the physics depends very much upon the boundary conditions. The detailed understanding of the statistics of the device becomes crucial. For example, if you make a very small device, you will have only one or a few atoms of impurities for the doping. This will strongly limit the application of statistics. This is a question of fundamental physics. I believe that the development of future technology will pose significant new physical questions for which we can't guess answers now.

This is why I am cautious, although I am an optimist in a statistical sense. It typically takes from 10 to 35 years for an invention to become a reality and only a small percent of the inventions in this field become useful products. My optimism is related to the following: there is already a quantum device that I have not discussed, called the quantum well laser. It was invented in the 1970s and is now a product 10 years later. Thus I am optimistic about quantum devices, but I am skeptical about trying to make conclusive statements about the physical limits of computation because the engineering pathway to the hardware may uncover new physical questions that had not been foreseen. For example, the discovery of the quantum Hall effect in heterojunctions was an amazing piece of physics that came out of work with semiconductors in a very unexpected way. There is a point in time that is optimum to ask a particular question, and I am not sure that point is now for your question.

*Drexler:*   I agree with the general direction of your remarks. I am trying to set lower bounds on what we can see as possible. I believe that in general we will discover things on the technological development path that will eventually make those predicted bounds seem silly, rather like the results that Babbage might have had if he'd tried to predict what Twentieth-century computing would be like.

*Capasso:*   We project that the physical limits to the speed of the bipolar transistor will be a transistor with a cutoff frequency of 500 GHz. Electrons in such a device will travel at velocities in excess of 10 cm/sec because the relevant device dimensions (base and collector widths) are comparable to the electron mean free path. The frequency of the device that our scientists built (500 Å base, 3000 Å collector, with an average velocity of $4 \times 10^7$) was 165 GHz, so that I think we have another factor of three or so to go. The optimists say that the limit is 1000 GHZ, which could be obtained if you achieve totally ballistic transport, as with a vacuum tube—an unlikely situation in a semiconductor.

*Ward:*   There is a class of compounds called mixed valence compounds that have been described as double harmonic well structures. What is the possibility that MBE can be used to make a double harmonic well structure?

*Capasso:*   We have made those structures and are now working on a superlattice structure made of parabolic wells. In another experiment some of our scientists are trying to measure quantum coherence in two coupled wells. In physics there is still a big controversy over the tunneling time. The experiment that we can do is the following. We have two energy levels in two coupled wells. We excite the electrons in one well only by tuning the energy of excitation. Quantum coherence then develops by tunneling, so that with properly designed structures we should be able to answer questions about quantum coherence.

# 9

## Fundamental Physical Constraints on the Computational Process

Norman Margolus

This chapter discusses fundamental constraints on computing machines that come from physics and some ways that computations can be reorganized to deal with these limits efficiently. These considerations are technology independent and apply at a suitably microscopic scale, such as where one is attempting to compute with small groups of atoms.

When discussing physical limits, it is not always clear which constraints are technological and which are fundamental. For example, is our present reliance on atoms as building blocks something that we can never transcend, or will we one day be doing computations with subatomic particles? In this chapter, I limit my discussion to those constraints that arise from general properties of physical law as we know it today, such as the finiteness of the speed of light and the reversibility of microscopic dynamics. Recognition of such constraints has motivated new models of the computational process that incorporate the constraints in order to allow them to be most efficiently dealt with.[1]

### What Is a Computer?

In 1986 I gave a talk at a conference on quantum measurement theory about quantum mechanical computers. I got involved in writing Hamiltonian equations, explaining why a parallel quantum computation had to be made asynchronous, and so forth. At the end, some brave soul asked me, "Yes, but what is a computer?" When you get into thinking about making a crystal of spins perform a quantum computation, you have to know what it is that you are trying to capture in your model, because it doesn't look like an ordinary computer.

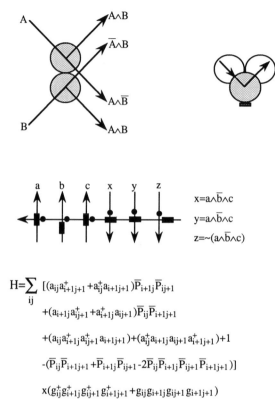

$$H=\sum_{ij} [(a_{ij}a^{+}_{i+1j+1}+a^{+}_{ij}a_{i+1j+1})\overline{P}_{i+1j}\overline{P}_{ij+1}$$

$$+(a_{i+1j}a^{+}_{ij+1}+a^{+}_{i+1j}a_{ij+1})\overline{P}_{ij}\overline{P}_{i+1j+1}$$

$$+(a_{ij}a^{+}_{i+1j}a^{+}_{ij+1}a_{i+1j+1})+(a^{+}_{ij}a_{i+1j}a_{ij+1}a^{+}_{i+1j+1})+1$$

$$-(\overline{P}_{ij}\overline{P}_{i+1j+1}+\overline{P}_{i+1j}\overline{P}_{ij+1}-2\overline{P}_{ij}\overline{P}_{i+1j}\overline{P}_{ij+1}\overline{P}_{i+1j+1})]$$

$$\times(g^{+}_{ij}g^{+}_{i+1j}g^{+}_{ij+1}g^{+}_{i+1j+1}+g_{ij}g_{i+1j}g_{ij+1}g_{i+1j+1})$$

**Figure 9.1**
Examples of computing elements (top to bottom): billiard balls, rod logic, Hamiltonian equation.

When you say the word "computer," the image that comes to mind for most people is probably a box with a CRT and a keyboard. But there are lots of other computing mechanisms one can think of. The most ubiquitous example of a computing element today is of course an integrated circuit. A variety of other, lesser known computing elements are shown in figure 9.1. For example, you can perform any computation using only billiard balls and reflectors. This is a very important theoretical result because it demonstrates that a classical mechanical system can do reversible computing. As a practical system it is not directly usable; you need perfect collisions, and so on. Another example of a classical mechanical computer is Drexler's rod-logic computer.[2] As an example of a very different computer, the Hamiltonian for a two-dimensional, universal, quantum cellular automaton, is "shown" at the bottom of the figure. This expression describes a rather simple computer in complete detail. The raising and lowering operators and projection operators totally prescribe the dynamics of the computational degrees of freedom. This equation is what got me into trouble at the meeting on quantum measurement; it is not obvious how a quantum automaton is similar to a "normal" computer.

What these devices all have in common, and in fact what we really mean by the word *computer*, is this: they are machines for transforming information.

We make all sorts of mechanisms to control the movement of matter and energy, but machines that deal only with information have a unique property—they do not have a characteristic size.[3] There is no point in miniaturizing an automobile to microscopic size because it would be too small to drive. But in a machine that transforms information, the bits do not have to be big; they can be as small as we are able to make them. Only their dynamics matters. It is quite reasonable to try to make computers very small, and people have been working hard at doing just that.

In old-fashioned adding machines, the positions of large mechanical parts represent digits in the computation. In a modern microprocessor, voltage levels in wires represent ones and zeros, and tiny complex arrangements of metal, silicon, and dopants are used to produce a desired sequence of logical transformations, such as those needed to add a set of numbers. In both cases, we have an abstract digital dynamics that we have managed to map onto the dynamics of the world. As physical laws make the parts of our machine move and change, we see the digits change in the desired manner.

Rather than creating a different machine each time we need to model a different digital dynamics, we have built general-purpose computers that can simulate *any* digital dynamics that can be realized by a machine, including other computers. Since they can simulate each other's operation exactly, all general purpose computers are equivalent. Except for issues of speed and memory capacity, all computers can perform the same set of information transformations. In fact, computer scientists believe that digital computers can perform any information transformation allowed by the laws of physics; this assumption, which has yet to be contradicted, is sometimes referred to as Church's thesis. Given the finite information-processing capacity of a finite system implied by the general properties of quantum mechanics, this thesis seems to be a good candidate for a fundamental law of physics.[4]

We characterize computers as universal information transformers, where information is a quantity represented by physical degrees of freedom on a scale whose minimum size is limited only by our ability to organize matter and energy on the smallest scales, and by any fundamental physical constraints on the minimum size of a bit (currently, none are known). Machines with a very simple regular structure can be used as computers by putting all of the structure and complexity of a problem into the program.[5] This simply recapitulates the process we go through when we build anything. Every machine is just a particular initial state for the operation of the laws of physics, which are uniform and regular.

The process of making computers on ever smaller scales, using physical degrees of freedom and the laws of physics in an ever more efficient manner, can be thought of as a refining process. In this process we are trying to refine and concentrate the computing power inherent in the laws of physics. You might ask, "Can we achieve 100% pure 'Computronium'?" From this point of view, the laws of physics are really the machine language, and any computer we build is just an initial state, it is just a program. If we want the most efficient program, we want to see how tight the code can be. That is, we want to see how close we can get to having the basic laws of physics compute for us in a direct way. We don't yet know the basic laws of physics so we cannot say too much about this now. But any attempt to bring our models of computation in line with the general constraints of our present conception of physical law can lead to significant benefits.

## Thermodynamic Constraints on Computation

Probably the most widely discussed limits on the computational process come from thermodynamics. It has become well known that there is an absolute lower limit on energy dissipation of something like $kT \ln 2$ per bit that is erased in a computation (where $kT$ is the typical thermal energy of a degree of freedom). Today's computers dissipate something like $10^{10}$ $kT$ per bit erased, so we are far from that limit. In the genetic apparatus, RNA polymerase dissipates about 20 $kT$ per base that it copies, which is much more efficient.[6] If our technology is so far from $kT$, why should we be worried?

The difficulty is essentially the surface-to-volume problem. Surface area varies as the square of the radius, while volume varies as the cube of the radius. This is the reason why animals cannot be made indefinitely large; they generate heat in proportion to their volume but only get rid of heat in proportion to their surface. We have exactly the same problem in computational devices. If we build a three-dimensional computer too large, it will melt. Either that or we have to put the pieces further apart, make the wires longer, and then the computer will be slower.

I want to present the idea that irreversibility (or noninvertibility) is the fundamental source of the energy dissipation. This notion that something like $kT$ per logical operation was needed had been folklore since the 1940s, but no one had carefully analyzed the requirement. Landauer presented a clear analysis in the 1960s.[7] Before this, it was not appreciated that it is the erasure of information that is costly.

One of the basic operations we want to perform with a computer is to clear a register. Most people cannot imagine carrying out a computation if we could not clear registers. This is clearly a logically irreversible operation: there is no way to know what the input was from the result. The NAND gate, a commonly used and universal logic element (with enough NAND gates you can build any logical function), is an irreversible operation. For three possible input values of A and B (00, 01, and 10), the output is 1. Thus if we only know that the result *is* 1, it is impossible to determine which value was the input. That is irreversible. All of our computers—as we build them today—are based on similar, logically irreversible operations.

However, the basic laws of mechanics are microscopically *reversible*. When a conventional "irreversible" computer clears a register, we cannot actually erase the information from the current state of the world (microscopically there must remain enough information to reverse the operation); we can only move the unwanted information out of the way. Therefore, in order to "throw away" a bit from the computational modes

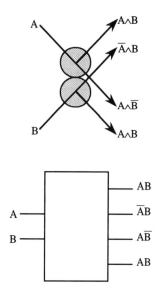

**Figure 9.2**
Reversible computation using a billiard ball computer.

(the degrees of freedom the computer uses to represent the bits of the computation), we have to double the number of states of the heat bath so it can represent an additional bit. To double the number of states of the heat bath, you have to add $kT \ln 2$. This is the source of the dissipation limit.

If you think about it carefully, this argument actually leads to a *constraint* on computations, but not to a limit. Theoretically we should be able to reformulate our computations to avoid this constraint. This was not obvious. Arguments such as the previous one—that clearing a register is irreversible—obscured the fact that you could avoid irreversibility in computation. One simple example, Fredkin's Billiard Ball Logic, helps us see how you can indeed have reversible computation (figure 9.2).[8] Imagine two paths, A and B, coming in from the left. Either a ball or no ball comes in on each of these paths. Each ball has a finite diameter. If B comes in and nothing comes in on A, a ball comes out on the path labeled "B and not A." If something comes in on both A and B, balls comes out on the paths "A and B." In this way colliding billiard balls can perform the AND and NOT logic functions. Such collisions,

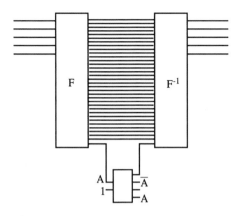

**Figure 9.3**
Scheme for avoiding dissipation in reversible computations.

with balls directed from one collision site to the next by suitably positioned ball-reflectors, can be used to construct any logic function: the
billiard ball collision is in fact a universal logic element. (A schematic
symbol for this "logic gate" is shown at the bottom of figure 9.2.)

These "collision gates" are clearly invertible. If the motions of all balls
are exactly reversed, a system of billiard balls and reflectors will run
backward and return to its original state. The only problem with these
reversible gates is that, when we use them to perform ordinary irreversible logic functions, such as AND or NOT, extra outputs are produced.
But it is these extra outputs that allow us to invert the overall function.
Reversible logic can perform any computation that conventional logic
can, but instead of dissipating information into the heat bath at every
logical step, as conventional logic does, reversible logic explicitly includes
extra information in the model to avoid any irreversibility. Eventually
we need to clear the computer in preparation for a new computation.
Have we gained anything by refusing to throw away all of this extra
information immediately?

Figure 9.3 shows how to "clear" the computer reversibly. Suppose
"F" performs a reversible logic function in which it takes in a five-bit
value, and, after a period of time, produces a one-bit answer, "A"—plus
a large number of intermediate results that comprise the extra information needed to go backwards. At this point, we could make a copy

of "A" in a reversible way (using a collision gate, as shown at the bottom of the diagram) and then send all of the intermediate results through a mirror-image (inverse) circuit "$F^{-1}$." Out comes the original five-bit value definition of the problem. All the other information is gone because it has been recombined. All the extra variables that were holding that information have been eliminated without doing any irreversible operations. In this way, the answer is obtained by performing "F." Then, by executing an extra phase of the computation, "$F^{-1}$," we clear the computer and return it to its original state. Thus the intermediate information is not dissipated at all, and, in principle, there is no dissipation proportional to the length of the computation but only proportional to the size of the question (assuming that we throw the question away once we have obtained the answer).

Of course this example is an idealized case. In reality we always have some dissipation proportional to the length of the computation, but this is a technological factor that can be made very small. There is no fundamental number that restricts the decrease of this dissipation. In fact, dissipation per gate can be far less than $kT$, and reasonable schemes for achieving this have been proposed using, for example, superconducting technology.[9]

Reversibility seems to be a very fundamental constraint. We do not expect this fact to change when the laws of physics are better understood. Essentially all physical theories depend very heavily upon reversibility: the unitary structure of quantum mechanics, the fact that heat engines cannot "erase" entropy, the fact in relativity that initial data on one hypersurface implies initial data on another hypersurface, part of which is "before" the first one. All of these crucially depend on the invertibility of the dynamics. Reversibility is the sort of constraint that produces conclusions that we expect will continue to be true despite any progress in physics or technology.

It is also important to understand that we need to know how to exploit reversibility in order to make microscopic quantum mechanical models—not only to avoid thermodynamic constraints but because quantum mechanics has invertible dynamics. If we could not even conceive how to perform a reversible computation, we would be forced to use only macroscopic computers.

## Quantum Mechanical Constraints on Computation

When we think about quantum mechanics, people tend to immediately invoke the Heisenberg uncertainty principle (the product of the uncertainty in the energy and the uncertainty in the time cannot be made arbitrarily small) as a constraint. The usual argument runs something like this: a computer should be fast, implying a short cycle time, implying a large uncertainty in energy, implying much dissipation. This argument is not correct. The error is that the energy uncertainty need not be dissipated. A completely autonomous quantum dynamics is possible, where everything is coupled coherently by quantum interactions. In this case the evolution of the system is governed by quantum Hamiltonian equations. The time uncertainty becomes only an uncertainty as to when the answer will be obtained. There is no actual constraint on how quickly the computation can be done. Of course, the rate of time evolution of the system depends upon the energy scale of the Hamiltonian. So, if we want a fast computation, we need strong coupling. But if the coupling occurs in a closed microscopic system, the energy is conserved rather than dissipated.

Feynman proposed doing quantum computation using a number of reversible logic gates connected by wires that are themselves reversible, exchanging what is at either end as signals travel down the wires.[10] Because a sort of "fuse" causes each gate to operate in turn, as the "burning" part passes by, only one gate operates at any one time. By expressing all variables as spins and implementing the fuse as a chain of spins with a spin wave on it, the computation can be carried along by the dynamics of the fuse. We can even make a wave packet state that runs at a uniform rate. The only uncertainty is simply the uncertainty about when the computation will be completed.

Feynman's model can be improved upon.[11] The Hamiltonian in figure 9.1 represents a quantum cellular automaton that performs a parallel computation. It turns out, however, that if we restrict the interaction of the spins comprising the system to a finite range, it is not possible to force all spins to update simultaneously. Some points are updated more often than others, leading to local variations in computational "time." The Hamiltonian in figure 9.1 describes a system with this sort of asynchronous parallel updating that nevertheless manages to perform ordinary deterministic computations using individual spins as "bits."

## Computation with Cellular Automata

One constraint already felt in current computer design is the speed of light. It takes a nanosecond for light to travel 30 cm; there is great motivation to make wires in a computer as short as possible. A model of computation that uses only short wires is a cellular automaton (CA), a uniform array of locally interconnected processing elements. Imagine a two- or three-dimensional array, such as a square grid, with a processing element at each point on the grid. Each processing element is connected only to its nearby neighbors.

This is exactly the sort of scheme people building VLSI would love to use. CA can run very fast because the wires are short. They can also scale up in size just by adding more elements, but without increasing the cycle time, because there is only local intercommunication. In some sense, it is an ultimate architecture. The question is, "What is it good for?" We don't know a priori that CA can do anything useful!

CA have been studied in the past mostly as a recreation (notably the game of Life[12]). The usefulness of CA is one of my major areas of research, and, when I talk about CA, I'm often asked if there are any interesting applications other than the game of Life. This seems very strange to me. It is as if someone who knew about differential equations thought the only interesting application was heat equations. There are lots of interesting CA, but we cannot easily study CA until we have machines built with CA hardware. The problem is that CA running on dedicated CA hardware run faster than software simulations by so many orders of magnitude, we cannot meaningfully simulate them on conventional computers. I have been involved, along with Tom Toffoli, in creating low-cost CA machines that already show a gain of a few orders of magnitude in speed. This preliminary success should encourage more exploration of CA. Our book describes the CAM-6 machine and a variety of physical simulations that we found it useful for.[13]

After inventing some reversible CA schemes, I got involved in making machines because I wanted to see what these reversible CA could do and I could not see much with ordinary computers. More and more of

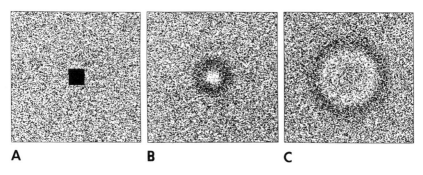

**Figure 9.4**
Simulation of a lattice gas on a cellular atomata machine.

my time is spent designing bigger and better CA machines—our current machine is the CAM-8.[14]

One interesting application of CA is illustrated in figure 9.4. This is a simulation of a lattice gas on a grid of 256 × 256 sites, in which particles move from site to site, moving in only four directions along the lattice and at one speed. In the figure, we start with a random distribution everywhere except for a block of material in the middle. The dynamical evolution of the system from panel (a) through (b) and (c) generates circular waves. With only slightly more complicated rules, we demonstrate ordinary fluid dynamics in two dimensions, at least for low speeds and for average particle flows.[15,16]

## Conclusions

In the same sense that theoretical chemistry is really physics, much of the study of theoretical computer science is also physics. This will become increasingly apparent as we continue to "refine the Computronium" that we find in nature. As models of computation are made closer to physics in order to make them deal more efficiently with fundamental physical constraints, the computations themselves begin to take on a more physicslike character. We have found that computations exhibit conservation laws and that reversibility guarantees that you have exactly the same constraint as the second law of thermodynamics inside your models. That is, no process of a fixed size that takes place within a system can indefinitely output less information than comes into it.

The following question comes to mind. Depending on how close the match is, on how well physics can be modeled in terms of information, is it possible that theoretical physics will turn out to be computer science?

## Notes

1. N. Margolus, *Physics-Like Models of Computation* (Cambridge, Massachusetts: MIT Press, forthcoming 1993).

2. K. E. Drexler, "Molecular machinery and molecular electronic devices," *Molecular Electronic Devices II,* ed. F. Carter (New York: Marcel Dekker, 1987, 549–571).

   K. E. Drexler, "Rod logic and thermal noise in the mechanical nanocomputer," *Proceedings of the Third International Symposium on Molecular Electronic Devices* (North Holland: Elsevier Science Publishers B. V., 1987).

   K. E. Drexler, "Nanotechnology and future supercomputing," *Proceedings of the Third International Conference on Supercomputing,* vol. 1, ed. S. Kartashev, S. Kartashev, International Supercomputing Institute, 1988, 512–516).

3. This characterization is due to Ed Fredkin.

4. See J. Beckenstein, "Entropy content and information flow in systems with limited energy," *Physical Review D.* 30 (1984): 1669–1679.

5. N. Margolus, "Physics-like models of computation," *Physica D.* 10 (1984): 81–95.

6. C. Bennett, "The thermodynamics of computation—a review," *International Journal of Theoretical Physics* 21 (1981): 905–940.

7. See R. Landauer, "Dissipation and noise immunity in computation and communication," *Nature* 335 (1988): 779–784.

8. E. Fredkin, T. Toffoli, "Conservative logic," *International Journal of Theoretical Physics* 21 (1981): 219–254.

9. K. Likharev, "Classical and quantum limitations on energy consumption in computation," *International Journal of Theoretical Physics* 21 (1981): 311–326.

10. R. Feynman, "Quantum mechanical computers," *Optics News* 11 (February 1985): 11–20.

11. N. Margolus, "Quantum computation," *Annals of the New York Academy of Sciences* 480 (January 1986): 487–497.

   N. Margolus, "Parallel quantum computation," *Complexity, Entropy, and the Physics of Information,* ed. W. Zurek, (Reading, Massachusetts: Addison-Wesley, 1990).

12. E. Berlekamp, J. Conway, R. Guy, *Winning Ways for Your Mathematical Plays,* (New York: Academic Press, 1982).

13. T. Toffoli, N. Margolus, *Cellular Automata Machines: A New Environment for Modeling* (Cambridge, Massachusetts: MIT Press, 1987).

14. N. Margolus, T. Toffoli, "Cellular automata machines," *Lattice Gas Methods for Partial Differential Equations,* ed. G. Doolen (Reading, Massachusetts: Addison-Wesley, 1990).

15. N. Margolus, T. Toffoli, G. Vichniac, "Cellular automata supercomputers for fluid dynamics processing," *Phys. Rev. Letters.* 56 (1986): 1694–1696.

16. B. Hasslacher, "Discrete fluids," *Los Alamos Science,* vol. 15 (special issue, 1987).

17. See G. Sai-Halasz, M. Wordeman, D. Kern, S. Rishton, E. Ganin, H. Ng, D. Moy, T. Chang, R. Dennard, "Inverter performance of deep-submicrometer MOSFETs," *IEEE Electron Device Letters,* 9 (December 1988): 12.

## Discussion

*Birge:* I am not sure that I understand your statement that quantum mechanics does not represent a constraint.

*Margolus:* It represents a constraint rather than a limit because it is a constraint that we can live with. Similarly, reversibility is a constraint, but it does not mean that we have to dissipate $kT$ per gate, it only means that we have to reorganize the way we do things.

*Birge:* So you are arguing that if you had a microscopically reversible system, you would never need to be able to get an answer before the next portion of the machine had to operate.

*Margolus:* Right. You don't have to meddle at each step with fat, macroscopic fingers between the logic gates. There is an interaction with the macroscopic world to initialize the computation, and again at the end to get the answer. In between it runs as an autonomous system according to the laws of quantum mechanics.

*Drexler:* If we do not meddle in between the gates, and if we make the physical assumption that we are building with atoms, and that the energy terms in the Hamiltonian that describe the quantum dynamics are limited by the kinds of energy we see in solid state physics, chemistry, and so on, could you comment on the limits for the mean switching speed inside the system?

*Margolus:* Assuming that each of the terms in the Hamiltonian is of the order of 0.1 electron volts, Feynman estimated that the switching time for his serial quantum computer would be on the order of $6 \times 10^{-15}$ seconds—6 femtoseconds. This is only about three orders of magnitude faster than switching times for recent silicon transistors.[17]

*Audience:* I am not sure of the size of your CAM-8. What sort of limits are you running into in terms of making these machines?

*Margolus:*   The fastest possible machine would be parallel at the level of individual cells. Because it is too hard to connect together billions of little cells, we have instead designed a machine composed of larger modules, each of which runs its own chunk of the space sequentially, and we time-share wires connecting cells in adjacent modules. We can build this architecture with ordinary serial memory chip technology, and it already gains a cost/performance factor of at least 1000 over existing, nondedicated machines. A typical, small machine has several million cells arrayed in three dimensions; larger machines can have billions of cells.

## Postscript

It might be reasonable to criticize this discussion as much too theoretical. I consider only perfect machines, without any noise from the outside world. Long coherence lengths and almost perfect shielding are certainly beyond our present technologies. Many researchers take the stance that noise and imperfections will always be the major constraints we must deal with at very small scales. From this viewpoint, bounds on perfect computing mechanisms are not very interesting; they may be wildly optimistic.

I am most interested in how constraints from physics force us to reorganize our computational schemes. I'm also interested in understanding physics from a computational point of view. So I have not been very concerned with noise and imperfections. In more practical conceptions of computers built at the molecular scale, using mechanisms similar to genetic machinery, logical reversibility and small size are still very much primary considerations.

Inasmuch as the world is quantum mechanical, every computer is a quantum computer. Distinctive quantum effects are exploited to make transistors work in every electronic computer today; these effects employ both averages over space (many copies of interacting quantum systems) and over time (many electrons pass through each system). Quantum interference devices, using less averaging over space, are getting progressively smaller. I see no reason in principal why we cannot fabricate much more complex components with relatively long coherence lengths.

This would mean fewer points of statistical interface between components and hence greater efficiency. Whether we will ever make effectively perfect machines isn't clear. However, I think we can certainly approach this limit, given the perfect, discrete nature of particles and quantum interactions in confined systems.

# 10

## Nanotechnology from a Micromachinist's Point of View

Joseph Mallon

I am neither a futurist nor a research scientist; rather, I am an engineer actively involved in making products. Therefore, my normal time frame is this quarter's results. My long-term planning horizon is normally five to ten years. However, I believe that nanotechnology is likely to affect the area in which I work (the micromachining of small structures), if not in the next five to ten years, then very likely within the next ten to twenty years.

I was introduced to nanotechnology when a friend pointed out A. K. Dewdney's article in *Scientific American*.[1] Then I read *Engines of Creation* and several related articles.[2]

My initial reactions were largely negative. I was skeptical of the validity of the arguments presented. Writing about nanotechnology seemed to be premature and overly speculative. It seemed too far in the future to warrant serious thought now. Practical tools suitable for implementing the technology obviously were not available.

But nanometer structures seemed at least *physically* possible. This struck me as particularly significant. The success of the scanning tunneling and the atomic force microscopes demonstrated that working with structures of atomic dimensions is feasible. Here was a clear connection between my familiar micro-scale world and the atomic- and molecular-scale world that Drexler was writing about. Microprobes for such devices could be made by micromachining techniques.

However, I felt that the present path to small devices had not been fully exploited. There is so much to exploit and so many problems to solve in micromachining, why worry now about devices orders of magnitude smaller? After further reflection, I realized that the overall trend

**Table 10.1**   Scale of micromachined devices

| Device | Conventional | Micromachined |
|---|---|---|
| Pressure sensor | $4 \times 10^{-6}$ m$^3$ | $1 \times 10^{-14}$ m$^3$ |
| Relay | $4 \times 10^{-7}$ m$^3$ | $1 \times 10^{-14}$ m$^3$ |
| Electric motor | $4 \times 10^{-7}$ m$^3$ | $5 \times 10^{-14}$ m$^3$ |

toward smaller and smaller devices seemed a logical extension of the efforts of the last several decades in silicon technology. Nanotechnology would (eventually) make compelling sense.

## Scale of Micromachined Devices

The devices presented in this chapter are rather large compared to atomic-scale devices. But they are useful devices that we can make today, and they are much smaller than comparable devices made with conventional techniques (table 10.1).

## Evolution of Micromachining

Micromachining has evolved over the last few decades, slowly at first, in a fragmented manner, then rapidly and with more focus. The technology is closely connected with solid-state sensors. Sensors are one of few commercially interesting silicon-based devices that require three-dimensional geometries. The early tools of micromachining were developed as tools for the manufacture of these devices. This development has very closely paralleled, and uses almost exclusively, the technology developed for silicon integrated circuits (ICs). In fact one of the most powerful aspects of micromachining technology is its ability to constantly draw on a well-researched and very successful body of IC technology.

The annual market for sensors is about a $4 to $5 billion. This is a substantial business, and a good fraction of the market has either converted or is in the process of converting to silicon sensors. Even so, investments in mainstream IC research and development are larger than the entire sensor business.

Micromachining is beginning to emphasize microstructures, but they are not yet commercially successful. However, significant devices that will have an important economic impact have been demonstrated in university and corporate laboratories. One of the great strengths of the field has been a continuous interaction between basic research and commercially successful products. Atypical of many other research areas, micromachining investigators at universities are strongly oriented toward making working devices. Consequently, the evolution of the field has been rapid—particularly in recent years.

Today, the most visible trend is toward making complex devices with moving parts. Five years ago, micromachining meant machining simple shapes and cutting grooves and holes in devices. Now we are making true machines.

### Engineering with Silicon

Micromachining is defined as "the three-dimensional sculpting of silicon using standard semiconductor processing." It is possible to micromachine many other materials, but the technology for working silicon is so exquisitely evolved that "micromachining" and "silicon" run together quite naturally as a single phrase. However, current deposition and epitaxial technologies used to manufacture mainstream ICs permit almost any engineering material to be used. In the future, we will see a wider variety of materials commonly used.

Micromachining technology was first developed in the 1960s and is today the technology of choice for making pressure sensors and accelerometers. It has enjoyed such success particularly because we can make many complex silicon devices simultaneously by employing batch processing.

A micromachined sensor is a three-dimensional solid-state structure for sensing physical variables. Silicon was initially employed as a sensing material primarily because it is highly sensitive to a variety of physical variables. While this mechanical-electronic property was the original reason for choosing it for sensors, silicon is also an excellent structural material.

Silicon is readily available commercially in a very pure form and has the following characteristics:

- High modulus of elasticity: $30 \times 10^6$ psi (same as steel)
- Low density: 2.3 g/cm$^3$ (same as aluminum)
- Reasonable hardness: 850 kg/mm (same as quartz)
- High melting point: 1400° C
- High tensile yield strength (stronger than steel)
- Essentially perfect elasticity
- High sensitivity to physical variables

Silicon structures can be very strong. A small silicon structure can easily have a yield strength of 180,000 pounds per square inch (psi). With careful sample preparation, a yield strength of 300,000 psi can be achieved. Furthermore, silicon has a low thermal expansion coefficient, which makes it very useful for making high-temperature and dimensionally stable structures.

## Micromachining Technology

The main technological factor driving the success of micromachined sensors and actuators is the fact that the price of computing power has been declining on a steep learning curve. Processing power has become essentially free in today's systems, compared to the systems of 20 years ago. Sensors and actuators have thus become the cost- and performance-limiting aspect of most intelligent, closed-loop control systems. This fact has spurred on interest in the field, and several major universities and corporate labs, in both the United States and Japan, are active in this area. The field is widely recognized to be commercially important; companies that make microprocessor-based systems need to efficiently incorporate massive amounts of mechanical information into their production to control processes and hardware. Most companies involved in making these systems are thus actively engaged in research.

Sensor technology tends to be about a decade behind mainstream IC technology. At NovaSensor we use four-inch wafers, compared to six- and even eight-inch wafers employed in the mainstream industry. But, because our devices are small, we can make a very large number of sensors from each wafer. A typical pressure sensor is 2.5 mm in diameter, and we are shifting to a standard size of 1.0 mm. Approximately 6500 dice per wafer times 25 wafers per batch are processed in a single cassette. This means that a unit production batch is approximately 150,000 sensors. These devices have yields of 85%—similar to bipolar ICs. In our rather small facility, which is about five years behind mainstream IC technology, we can easily generate the world's supply of sensors (currently about 25 million sensors a year).

Micromachining processes are either additive or subtractive. Material is deposited on the wafer or removed in a selective manner using processes employed in conventional IC manufacturing.

The following are subtractive processes:

• Chemical etching (isotropic and anisotropic)
• Reactive and nonreactive plasma etching (removing material by ion bombardment)
• Electrochemical etching
• Mechanical grinding, polishing, and sawing
• Laser and ultrasonic drilling

The following are additive processes:

• Epitaxy (growth of material from the vapor phase)
• Thermomigration (thermally added diffusion)
• Field-assisted thermal bonding (wafer fusion with heat and an electric field)
• Polysilicon deposition (deposition of polycrystalline silicon)

Micromachined structures employ an additional degree of freedom compared to conventional ICs. A silicon IC is essentially a flat (two-dimensional) device. The device is incorporated in the top one or two microns of material. Micromachining techniques, however, allow us to sculpt both sides of a wafer. Using this additional dimension of freedom, we can fabricate various simple primitives and combine them together to manufacture rather complex mechanical devices and even simple mechanical systems.

Typical primitives are

• Grooves, cavities
• Shaped apertures: nozzles, mesas, orifices
• Beams, frames, bridges, needles
• Diaphragms, membranes
• Bimorphs (composites of two materials)

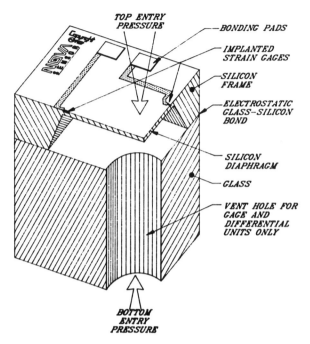

**Figure 10.1**
Diagram of a micromachined silicon pressure sensor.

Recently, we have learned how to make even more complex structures by laminating wafers together. We can fuse silicon wafers by polishing them, placing them in contact, and then heating to 1000° C. Using this technique, we can laminate several layers together to make closed silicon cavities and other structures.

A variety of sensing mechanisms are exploited to make silicon sensors, including variable capacitance and piezoresistance (change in resistance with applied stress), as well as vibrating, acoustic, and optical structures. Other sensors are based on thermal or ion transfer.

### Micromachined Pressure Sensors

A typical silicon pressure sensor is shown in figure 10.1. It has a rather thin membrane, usually on the order of 10 to 20 microns. A piezoresistor on the edge of the device senses the bending of the membrane under an applied pressure. As can be seen in the illustration, the sensor is stress isolated from its substrate by mounting it on a thick glass pedestal. The sensor is 2.5 mm square; the sensing diaphragm is 1.6 mm square. Anisotropic etching is used to create well-controlled dimensions. Plate 19 is a transmission micrograph of the device drawn in figure 10.1. Looking through the device, it is possible to see the four edge piezoresistors. The eight-micron-thick silicon membrane is reddish because silicon is selectively transparent to visible light, rejecting blue and transmitting red. Membranes as thin as 1.5 microns can be manufactured. These dimensions are large compared to the nanometer scale of molecular devices, but much smaller than conventional, last generation electromechanical sensors.

A typical application for these sensors is a disposable blood pressure sensor used in hospitals. Six to eight million such devices are used each year to monitor the blood pressure of postintensive care cardiac patients. The sensor is connected to the patient with a disposable fluid-filled catheter. Using similar technology we can make even smaller structures. A similar device, 0.4 mm by 1.0 mm (roughly the size of a crystal of table salt), is small enough to use in the blood vessels of the human heart to measure cardiodynamic pressures (figure 10.2).

Silicon fusion bonding can be used to create simple cavities (figure 10.3). The bond between the silicon wafers is of such quality that we can build active electronic devices through the interface. The thickness of the membrane sealing the top of the cavity can be controlled down to two microns.

There are two driving forces to make smaller and smaller sensors. First, smaller devices can fit into smaller places. Second, smaller devices are less costly to manufacture. The manufacturing costs of a device decreases as size decreases down to about one millimeter, at which point the price of the device is (currently) at a minimum. Smaller devices are more expensive due to yield, process complexity, and other reasons.

Obviously, the size of a finished device may have little to do with the size of its sensing element but is instead determined by other factors. For example, the sensing element of a manifold pressure sensor for an automobile is about one millimeter across, but the package is much larger. In 1988, several manufacturers made about 17 million pressure-sensing devices worldwide. Production is currently increasing at about 10% a year. At these quantities, it is possible, but generally not econom-

**Figure 10.2**
Micromachined silicon pressure sensor suitable for in vivo implant in human heart (shown with table salt crystals).

ically feasible, to put the electronics and intelligence on the sensing chip. It is more cost effective to process the sensor separately from its mating electronics and interconnect them at the package level using hybrid microelectronics.

Furthermore, the overall size of a device is often determined by the need for a human interface. If a mechanic needs to be able to install the device in an automobile, it needs a human-scale package and connector. This consideration may establish a principle for nanometer-scale devices for which there is a need for real world interactions: the requirements of a user interface may determine useful size. Micromachining may provide this interface because it produces devices between the nanometer and human scales.

## Other Applications of Micromachining

Applications of micromachining techniques that have either been demonstrated or are in limited commercial use include scanning tunneling

**Figure 10.3**
Closed cavity formed by silicon fusion bonding.

and atomic force microscopes, valves and switches, relays, fuses for detonating weapons, coolers, optical interface devices, and very thin membranes for use as electron or X-ray masks for next generation IC photolithography.[3]

It is even possible to make very small vacuum tubes. Why make vacuum tubes in the age of integrated, solid-state circuits? Electrons travel faster in a vacuum than in solids. Furthermore, vacuum tubes are insensitive to radiation. The problem with vacuum tubes was not their physics but their size, heat generation, and the inadequacy of the available manufacturing techniques. Imagine building a vacuum tube using micron-scale structures with a cold field emitter on a silicon wafer that was able to interface directly with modern digital electronics. Such a device is potentially very fast and radiation insensitive. A number of groups are pursuing this potential.

Figure 10.4 shows a device made by Kahil Najafi of the University of Michigan. This device is an implantable neural probe used to measure signals in the human brain. The active electrodes of the device are at the

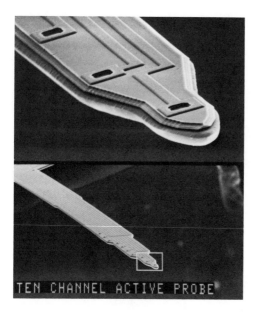

**Figure 10.4**
Neural probe. (Courtesy K. Najafi, Center for Integrated Sensors and Circuits, University of Michigan.)

tip of the silicon needle. The electrode regions, separated by approximately 20 microns, provide the ability to measure side-by-side neural signals in the brain, a feat impossible with existing electrode technology. Three of the probes are combined into a single structure, with added electronics to multiplex and amplify neural signals to generate high-level signals for analysis.

A device fabricated at NovaSensor is an ink jet nozzle with a small, well-controlled orifice for dispensing fluid, as part of an ink jet printer, for example (figure 10.5). The nozzle is created by etching a cavity the shape of a truncated pyramid in silicon and then forming a thin dielectric layer with a well-defined round hole to provide a 2.5-micron aperture at the top of the pyramid.

Another device developed at NovaSensor is a pressure regulator that works by magnetically driving a membrane in close proximity to a small hole (four to six microns) regulating the flow of air through the aperture (figure 10.6). This structure serves as an electrically driven pilot valve for use in precision regulators.

**Figure 10.5**
Silicon ink jet nozzle.

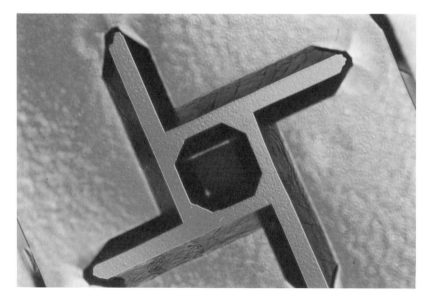

**Figure 10.6**
Orifice for small pressure regulator.

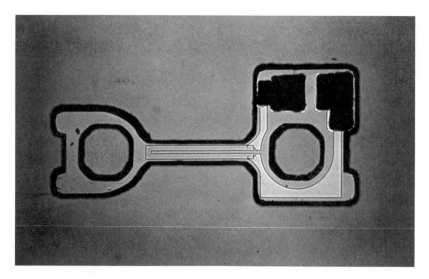

**Figure 10.7**
Silicon strain gauge for measuring heart muscle strain. (Courtesy T. Nunn, Kulite Semiconductor, Inc.)

A strain gauge fabricated by Tim Nunn, while at Stanford University, measures the strain in heart muscles. It is about 0.75 mm in length and includes holes for surgically sewing it into the human heart (figure 10.7).

A flow sensor developed by my colleague Kurt Petersen (while with his previous employer, Transensory Devices, Inc.) is a thermally isolated structure on a cavity-bridging dielectric structure of less than one micron (figure 10.8). The sensor itself is a thin-film resistor that heats up when a current flows through it. Then the resistor cools in proportion to the mass of air that flows across its surface. The device is used to measure gas flows in semiconductor processing. Early versions of this device failed because of locked-in stresses in the thin film. To minimize this and other performance problems, we used extensive computer modeling to study stress and thermal and electrical effects.

Barth and Zdeblick at Stanford University developed a silicon fluidic amplifier, with dimensions as small as 10 microns. Their device, which can be interfaced with electronics, demonstrated a gain of 5 to 10 and a calculated frequency response in excess of 100 kilohertz (figure 10.9).

It is also possible to make small silicon devices that move, although this work is in an early stage. Figure 10.10 shows a device developed

**Figure 10.8**
Silicon microflow sensor.

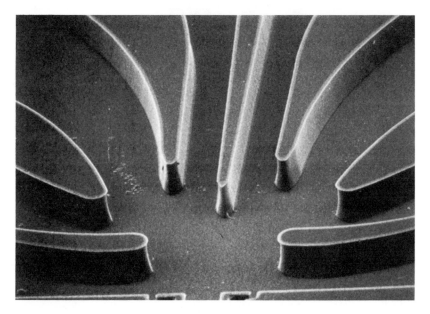

**Figure 10.9**
Silicon fluidic amplifier. (Courtesy M. Zdeblick, Redwood Microsystems, Inc.)

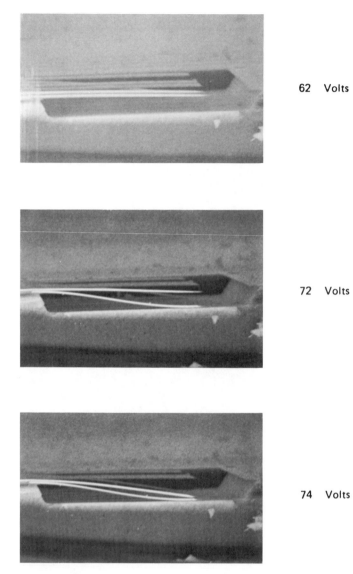

62   Volts

72   Volts

74   Volts

**Figure 10.10**
Silicon beam bent under applied electrostatic force.

by Kurt Petersen while at IBM. A small suspended insulator beam is progressively bent by applying electrostatic voltages between the silicon beam and the underlying substrate.

Such a device can form the basis of a relay. What advantages would a silicon relay have over other switching techniques? Solid-state switches, while they are fast, do not have infinite resistance when open nor zero resistance when closed. Mechanical relays are currently used in precision data-acquisition applications, where low millivolt signals are very important. However, these mechanical relays are large and expensive. Moreover, their manufacture is not easily automated. It is possible to make micromachined silicon relays with dimensions on the order of tens of microns that are very fast. The contacts of these relays are approximately 25 microns. We could extend this technology down to feature sizes of one micron, and perhaps to 0.5 or 0.25 microns. A current generation relay of about 40 microns is shown in figure 10.11.

Fusion bonding can be used to create a suspended structure in silicon. The width of the double cantilevered portion in figure 10.12 is about four microns. The dimensional control is excellent, and we have not yet approached the limits of this technology. However, limited effort is going toward making devices much smaller; the technology is relatively new and possibilities in the current size range, which is already smaller than conventional mechanical devices, are still underexploited.

An accelerometer is nothing more than an inertial mass suspended from a flexure with some means of detecting the deflection. The accelerometer in figure 10.13 detects this deflection by measuring the stress at the surface using a piezoresistor. One problem with a silicon accelerometer is that its range may be one G, but, if dropped on the floor, it can be exposed to 1000 Gs, or more. The device shown in figure 10.14 uses five-micron stops to achieve durability. The device also incorporates a narrow gap behind the inertial mass to keep the flexure from oscillating in a free manner. As the beam moves, air is forced through the gap, damping oscillations. The finished accelerometer, smaller than a grain of rice, is a complex mechanical structure embodying a mass, a flexure, and stops.

**Figure 10.11**
Silicon microrelay with gold-plated contacts.

**Figure 10.12**
Suspended silicon structure.

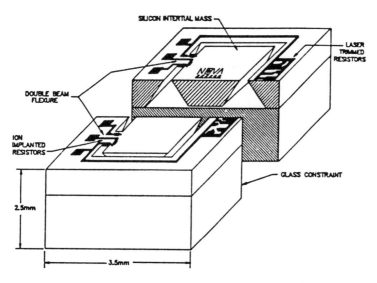

**Figure 10.13**
Schematic for micromachined silicon accelerometer.

Figure 10.15 shows another accelerometer, developed by Peterson while at IBM, that is smaller than the accelerometer shown in figure 10.14. The larger device was made to meet commercial specifications, whereas this device was made to demonstrate the capability of making high-density devices in a small space. In this particular case, the inertial masses are small bumps of gold made by electroplating.

Roger Howe, while working on his Ph.D. at the University of California, Berkeley, constructed a suspended polycrystalline silicon beam that vibrates with a resonant frequency that can be used to detect the presence of chemicals by selective adsorption on the surface (figure 10.16).

It is also possible to manufacture small cantilevered structures with dimensions on the order of a few microns. These structures can be deflected in the presence of a high electrostatic field. Similar structures can be used to make vibration monitoring devices that analyze the frequency of a vibration. Each beam has a different resonant frequency. Motion is detected by measuring the capacitance between the structure and the substrate. The output from each beam is proportional to the amplitude of vibration at its resonant frequency.

**Figure 10.14**
Micromachined silicon accelerometer with poppy seed.

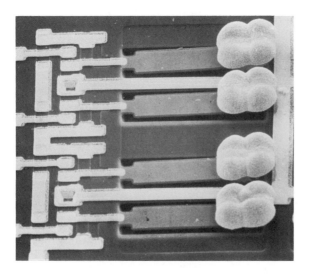

**Figure 10.15**
Silicon accelerometer with gold-plated inertial masses.

**Figure 10.16**
Polycrystalline silicon beam used in vibration mode as a chemical sensor.
(Courtesy R. Howe, University of California, Berkeley.)

## Micromachining Complex Machines

Stephen Terry, when he was at Stanford University, developed a gas chromatograph on a single silicon wafer, with all essential components of such an instrument including a solenoid, a microvalve that can dispense 100 nanoliters at a time, a capillary column, and a thermal detector (figure 10.17).

Another very intriguing device with a high level of system complexity is a commercial device made of glass, rather than silicon, that was developed by William Little at Stanford University. The device is a Joule-Thomson microminiature refrigerator, a portion of which is shown in

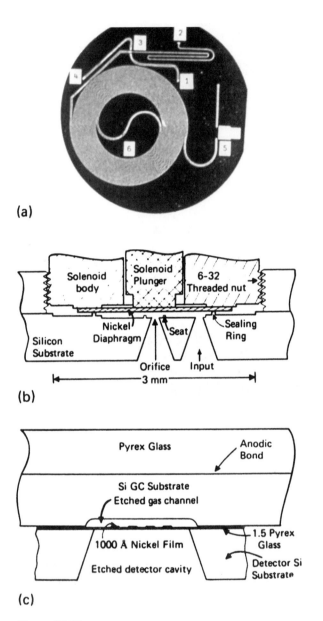

(a)

(b)

(c)

**Figure 10.17**
Schematic for gas chromatograph constructed on a silicon wafer. (Courtesy S. Terry, IC Sensor, Inc., from J. Angell, J. Jerman, S. Terry, S. Saadat, "A Proto-type gas analysis system using a miniature gas chromatograph," U.S. Dept. of Health and Human Services, Contract No. 210-77-0159, April 1981.)

figure 10.18. High-pressure nitrogen is introduced through an expansion nozzle. The device is used to cool microscope slides to cryogenic temperatures. Another cooling device developed at Stanford passes liquid nitrogen through small channels etched into the back of an IC chip (figure 10.19). Conventional cooling techniques can only cool chips at power dissipation levels of six to seven watts per square centimeter. Passing liquid nitrogen through channels behind the chip can dissipate as much as 2000 watts per square centimeter. Computational limits are ultimately established by the ability to dissipate heat; these cooling systems hold the potential for significantly extending these limits.

Work with devices that are actually machines is exemplified by electrostatically actuated tweezers developed by Noel MacDonald (figure 10.20). The tongs open and close, with the appropriate applied voltage, over a distance of about 10 microns.

Another interesting device is an air turbine developed by William Trimmer (figure 10.21). This device is created using a "sacrificial layer"

**Figure 10.18**
Photomicrograph of lower part of microminiature heat exchanger and expansion capillary abrasively etched in glass plate. (Courtesy W. Little. Reprinted with permission from W. Little, "Microminiature Refrigeration," *Rev. Sci. Instrum.* 55 (1984): 661–680. Copyright 1984 Am. Inst. of Physics.)

**Figure 10.19**
Micro-cooling channels for silicon IC. (Courtesy D. Tuckerman, nChip, Inc.)

technique in which a layer that initially bonds parts of the bearing together is removed. Silicon turbines have been spun at speeds as high as 500,000 rpm. An advantage of such a small device is that the tips of the rotor do not go supersonic at these speeds, as would the tips of a larger turbine. On the other hand, these devices can be excellent generators of silicon dust. Accordingly, the problem of good frictionless bearings is being vigorously addressed. Air and electrostatic bearings are being considered as are techniques for reducing friction at the surface, such as thin diamond films.

A torsionally resonant structure developed by Roger Howe at the University of California, Berkeley, is shown in figure 10.22. When stressed, this device, designed to resonate via electrostatic attraction between the tines, measures changes in dimensional length.

A rather exciting development announced first at the University of California and then at MIT is the development of silicon electric motors with diameters of about 100 microns (plate 20).

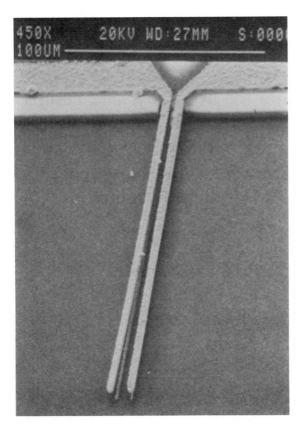

**Figure 10.20**
Micromachined silicon tweezers. (Courtesy N. MacDonald, Cornell University.
Reprinted with permission from N. MacDonald, L. Chen, J. Yao, Z. Zhang, J.
McMillan, D. Thomas, K. Haselton, "Selective chemical vapor deposition of
tungsten for microelectromechanical structures," *Sensors and Actuators* 20
(1989): 123–133.)

## Micromachining and Nanotechnology

How does all of this relate to nanotechnology? Micromachining is a
demonstrated, commercially viable technique for working on a scale that
is only a few orders of magnitude larger than molecular devices. The
relationship between these two technologies seems similar to the relationship between vacuum tubes and transistors. There was little connection between tubes and transistors—and perhaps similar discontinuity

**Figure 10.21**
Micromachined air turbine. (Courtesy W. Trimmer, Princeton University. Reprinted with permission from M. Mehregany, K. Gabriel, W. Trimmer, "Integrated fabrication of polysilicon mechanisms." *IEEE Trans. Elec. Dev.,* 35 (1988) 719–723. Copyright 1988 IEEE.)

**Figure 10.22**
Torsionally resonant silicon beam. (Courtesy R. Howe, University of California, Berkeley.)

between micromachining and nanotechnology—because of vast differences in fabrication techniques. On the other hand, successful silicon micromachining has created an atmosphere of acceptance for small mechanical devices. Fifteen years ago there was not a large body of people designing microscale structures. Now there is.

In addition, micromachining provides a possible interface technology. The scale of such devices is intermediate between the scales of the molecular and macroscopic worlds. Micromachining could also be used as a prototyping technology for structures intended for a smaller scale, just as engineers prototype IC chips by first developing a system using individual components and PC technology. Finally, micron-scale devices might provide an appropriate packaging technology for certain nanometer-scale devices.

In conclusion, it seems worthwhile to pursue the relationship between the two technologies, perhaps by developing a demonstration structure on an intermediate scale. Such a mechanical structure, with multiple elements and low but nontrivial complexity, could have large features of 1 to 5 microns and small features of 0.25 microns. This device should incorporate some degree of mechanical logic and be mobile.

In summary, I am excited by the possibilities. I do not know what the future holds, but I look forward to the developments of the next few years.

## Notes

1. A. K. Dewdney, "Nanotechnology: wherein molecular computers control tiny circulatory submarines," *Scientific American* 258 (January 1988): 100–103.

2. K. E. Drexler, *Engines of Creation* (Garden City, New York: Anchor Press/Doubleday, 1986).

3. Brief overviews of micro-manufacturing include: I. Amato, "Small things considered: Scientists craft machines that seem impossibly tiny," *Science News* 136 (1 July 1989): 8–10.

Also see P. Elmer-DeWitt, "The incredible shrinking machine: breakthroughs in miniaturization could lead to robots the size of a flea," *Time* (20 November 1989) 108, 110, 112.

## Discussion

*Audience:*   What solutions might be possible to the problem of friction in moving structures?

*Mallon:*   This is clearly a serious problem. Two approaches seem reasonable. One is to separate the surfaces by a suspension technique—such as an air or electrostatic bearing. The other possibility is to combine very smooth surfaces with a film of diamond, which has about the same coefficient of friction as Teflon in addition to being very hard.

*Audience:*   What is the typical intellectual background of someone working in this field?

*Mallon:*   I'm happy to say that it is now possible to pursue a degree in this field because a number of universities are working in the area. Typically, people are in electrical engineering departments because they use the tools of semiconductor processing, but they could also be found in mechanical engineering or in interdisciplinary programs. Most of the people I know are electrical engineers or solid-state physicists.

*Audience:*   I understand that there is a device that is in some use now that is a telemetering thermometer that people actually swallow. Was micromachining used to develop that particular device?

*Mallon:*   Yes. There is also a pressure sensor. There is also strong interest in esophageal pressures which measure the contractions of the digestive system. To my knowledge, these devices have only been demonstrated and are not yet in use clinically, but the technology is available.

# 11

## What Major Problems Need to Be Overcome to Design and Build Molecular Systems?

*Panel Discussion:* K. Eric Drexler, John Foster, Tracy Handel, Ralph Merkle (moderator), and Michael Ward

*Merkle:* This panel discussion will focus on some of the interesting problems raised by the rest of the conference—in particular, what are the major difficulties to be overcome in designing and building molecular systems? Each panelist will make a short statement, then respond to the other panelists' statements, then we'll invite the audience to participate.

*Drexler:* My paper addresses this issue directly, so I will simply comment that the fundamental challenge of nanotechnology is how to make things larger. We have worked up to controlling the structure of substantial macromolecules with protein engineering. The core question is how to build larger structures while maintaining detailed molecular control.

*Handel:* I have a relatively optimistic view of nanotechnology in terms of protein design. The database of known natural proteins and their folding motifs is increasing rapidly. This is largely due to advances in molecular biology and theoretical studies. For simple systems, the prospects of knowing how to make what we want to make look very good. An interesting question is how many subunits can fold together. It may be possible to use a polymer as a matrix to initiate folding for subunits of protein molecules, or we may be able to fold several different subunits together in a crystallization process.

The second challenge is to develop our tools. Again, I think this looks pretty good. We can make almost any protein that we want; the difficulty is getting them to fold together correctly inside cells. This issue is being addressed. For example, chaperonins are a hot topic.[1] These are proteins that are thought to guide protein folding. The problem of protein folding may be answered in the near future.

The area of nongenetically coded amino acids also looks promising. Solid-phase synthesis is currently used to create these, but some groups are trying to incorporate them genetically.[2] Considering other polymers, there's at least one example of a yeast that produces a polyester. So perhaps we can manipulate cells genetically to produce polymer-protein complexes. This is way in the future, but it certainly seems feasible.

One of the biggest challenges, at least in the near term, is characterizing what we do make. Right now it takes a long time to establish the structure of a protein. You want to be able to put something together in the lab, or induce cells to turn out a protein, drop it into an nuclear magnetic resonance (NMR) tube, and ten minutes later have a spectrum; but you can't do that. I'm sure John Foster could describe several techniques for characterizing molecular surfaces, but with proteins, you are essentially restricted to X-ray crystallography and NMR. Both of these take a long time, and with NMR, the maximum size of a protein that can be studied is limited. This constraint is easing as people begin to look into the third dimension. Magic angle spinning techniques are being used for solids, but I don't know how soon this will be useful. This is the major problem, speeding up the cycle of design, synthesis, characterization, and redesign.

*Ward:*   While listening to Drexler's talk, I recalled the story of a physicist who proposed to put a satellite into geosynchronous orbit, tether it to the ground, and use it as a launching and retrieval vehicle for cargo shuttles to the moon. When the physicist was asked how he would do this, he said "That's easy. All the the chemists and engineers have to do is construct a single crystal of Kevlar—22,000 miles long." This shows a cultural difference that we need to bridge. Chemists and engineers may be realists, while physicists tell us what is possible. Chemists have traditionally moved toward studying self-assembly, starting with covalent linkages. In a sense, nylon at DuPont was a good example of that. Historically, they've been looking at very strong interactions. Now they need to pay attention to weak interactions in order to understand self-assembly processes.

To a large extent, computer-aided design will help. It may help us understand weak interactions and eventually allow us to put pieces together and understand reaction pathways. Computer-aided design will help not only in looking at thermodynamics but also in studying reaction pathways, transition state theory, the energetics of activated complexes, and so on.

Therein lies a real need for cultural change. Chemists tend to have a background in activation theory, but they tend also to systematize every search and to look at things by functional groups or by systematic reactivity patterns. Perhaps they need to probe the fundamental aspects of the chemistry a little more and think about how to design systems.

In terms of what is possible chemically and what needs to be done, I agree with Dr. Handel that characterization of subassemblies on a scale other than what is typically done is going to be very important. That is, finding the structures of aggregates and probing structure-function relationships is going to be very important in order to understand what small molecular assemblies mean. Superimposed on this effort is a funding question. Obviously, we have a long-term goal—early twenty-first century—and we need to pursue funding for that research. But we may need to make incremental improvements as well. Considering the paper presented by Joe Mallon on micromachined systems, maybe we should construct a prototype, microscale rod logic computer to convince people that this is interesting technology.

*Foster:* I agree with everything presented, so I can't talk about that, but my perspective is a little different. I notice that I'm the only one here who doesn't know any chemistry. The thing that astounds me about chemistry is how chemists tend to think in three dimensions. Physicists don't think in three dimensions; surfaces are about as complicated as we normally handle; and one dimension will usually do.

Creating something on a surface—like some small rods that move and perform some simple computing—would be a major hurdle I'd like to see us overcome. The biggest challenge is simply getting people moving and doing something in this arena. If some intermediate solutions are technologically extremely valuable, in areas other than nanotechnology, those solutions will be real winners. If you could, for example, assemble nice single crystals of diamond, that would be a winner. People would get very excited and the funding would pour in. Or if you could look at a surface, that may or may not be a single crystal, and may or may not be something you made, but if you could fluorinate the surface by putting fluorines down just where you want them, you could make a van der Waals bearing surface. Such a surface doesn't have to be made on the nanoscale to be useful; it could be made on a macroscale and run in a Datsun. A surface that is fully fluorinated would be great to slide things on, and eliminating friction would be quite a hit throughout the world. Perfectly tacking fluorines everywhere on a single crystal surface would prove the feasibility of nanotechnology, in a crude way, and at the same time provide the kind of intermediate solution needed to continue.

*Merkle:*    The degree of agreement among the panel is really astonishing. I invite comments from panelists who would like to comment on statements of other panelists.

*Drexler:*    I have some disagreements. I think the panelists have been remarkably uncritical of the last presentation made this morning (that is, my own presentation). The presenter of that paper spoke about reducing friction, and he may have omitted to say that he was speaking of static friction as opposed to dynamic friction. In dynamic friction, there are issues of phonon drag, which are quite substantial—although fluorinated surfaces are indeed of considerable technological significance.

Also, I think that people are too fond of mechanical nanocomputers. That is a fine design exercise, but attempting to implement something like that in the short term is very foolish.

My general observation is that we should focus on tool development. In particular, I have a comment on the molecular characterization problem, tightening up the design-feedback loop. One attractive feature of nonstandard amino acids that I didn't address in my paper is that we should be able to make structures that are not merely more stable than those found in nature but also stable at smaller sizes. Proteins have a certain threshold size before the folding is stable. If we can push that threshold size down, by making the interactions stronger, we can create smaller molecular objects, which, if they are quite small and quite stable, will be relatively easy to characterize with NMR techniques. These small molecules would be relatively stable building blocks which could be used to build larger structures. This may be a very substantial help in the design-feedback loop.

*Merkle:*    Would any of the other panelists like to respond? If not, I welcome the audience to ask questions.

*Audience:*   This is a comment, not a question. I think that as we proceed toward nanotechnology, we will have to rely more on modeling. It will be impossible in the future to use today's analytical tools on a regular basis. We need computers that can provide reliable molecular structure and structure-function relationships. These will at first be based upon experimental data, but I don't believe that it will be possible to meet these needs in the future by analyzing the structures of large numbers of proteins (with NMR and X-ray crystallography). Thus, modeling, computer capability, and a national commitment to these, should all be major issues. This is a very serious political issue: the repercussions in industry due to bottom-line thinking and the reduction of commitments to research. How can we think about nanotechnology in the political environment that we have in this country?

*Merkle:*   Any comments from the panel on this?

*Handel:*   I was going to mention that we can avoid a lot of the problems with characterization if we know in advance that our predictions are going to be very accurate. But even if we could characterize large proteins with NMR, a 600 MHz NMR costs $1.5 million and is not something you find in every laboratory. With those financial costs, that kind of characterization is not feasible on a regular basis.

*Ward:*   I'd like to address that somewhat peripherally. Computers will certainly play a big role in design and in mapping out profiles of reactivity for self-assembly. But we need to get more computers into the hands of actual users. Also, synthetic organic chemists still find a large barrier to getting into computer modeling. We can solve this big problem in either of two ways: we can change them or we can change the computers. The computers are already being changed, and there are many advances being made to make them more user friendly.

*Foster:*   I think that it is time for somebody to be disagreeable since we have all been so agreeable. While I agree that modeling is very important, the business of developing tools other than computer tools is also extremely important. If I could disagree harshly with Bill Joy, it would be on his comments about the amount of modeling that is needed and how much it would cost, particularly when he compared those costs with the Superconducting Super Collider. While I don't know enough about the SSC, I do know that it is very expensive. He seemed to be saying that if we had enough computers, we could just model the whole thing, compare all the theories, and not need to do the experiment. I think everyone here realizes that that is completely ludicrous. As soon as you think you understand all the physics or chemistry that goes into a problem, and you think that you can just model it, you've just lost. You've just lost the game because someone else is going to do a fairly simple experiment, figure out something that you never thought of, and you will be proven dead wrong. Suddenly you find that you have been modeling the hell out of something that doesn't have anything to do with reality.

This point is very important. I also think that it is worthwhile noting that all of the Nobel Prizes that we have had lately, at least the ones I am familiar with, have come from very good ideas—sometimes just outrageous ideas—with almost nothing to do with modeling. High Tc superconductors, for example, are very impressive and are not going to come out of a computer model. Developing room temperature superconductors will require many experiments, at least during the next fifty years until nanotechnology arrives. We have to do the experiments to develop the tools so that we can do even more. And that's my pitch for everything—except computer modeling!

*Audience:* I would like to disagree violently with a point that was brought up just a few minutes ago that it would be nice if we could avoid all of this characterization "mess." That point of view sounds like, "We need to be able to design nanomachines, we need to be able to build them, but that we don't need to be able to debug them." This sounds very strange to a computer scientist.

*Drexler:* Debugging is certainly very important. I would like to point out that the focus of this panel is on molecular systems engineering. If someone has a real killer objection to advanced nanotechnology, please speak up, but the focus, I think, should be on pathways.

I'd like to suggest how one might be able to use tool development to do a better job of characterization. This is a rather ambitious intermediate goal: Learn to make molecular objects and get pretty good at it. Learn to make associations of them so that you can make fairly large structures by self-assembly. Make a box that is large enough to contain a typical soluble protein molecule of interest (membrane proteins are, as usual, more difficult). Include some ability to bind the protein to the inside of the box in an oriented way. Design the outer surface of the box so that the box can be readily crystallized into good X-ray diffraction quality crystals. Now you have a systematic methodology for taking your new modest-sized protein, or other molecule, putting it into a box, crystallizing the box that contains the oriented molecule, and doing X-ray crystallography on the result. A similar approach has been suggested using DNA to make a framework to contain proteins.[3]

*Merkle:*   Other questions? Any killer objections?

*Audience:*   To a large extent, this conference seems to have ignored a top-down approach to achieving nanotechnology. More than eighty percent of the conference has concentrated on coming up from the bottom through chemistry. This may be a mistake, because coming from the top down, as with STM equipment, you can see what you are doing, while coming from the bottom up, you must do a lot of theorizing about what you are doing. If nothing else, the top-down approach provides the visualization instruments for analyzing what's going on.

*Drexler:*   A brief semantic point: I would count STM- or AFM-based instrumentation as a bottom-up approach in that the challenge is to make larger structures. We have already seen single molecules pinned down. The first challenge is greater control, then larger scale construction. Nonetheless, the observation that you have a direct way of seeing what you are doing is significant.

*Foster:*   Yes, most of the emphasis has been on the chemistry approach—whatever you call it. What is most appealing about that approach is the huge parallelism (making trillions of objects simultaneously) and the fact that biology works. At the same time, since I work with the STM, I know how hard it is. Meanwhile I get the feeling that people who work with proteins know what they are doing, and that it all works pretty well. Perhaps the chemists will come up with the solutions and the physicists will plod behind.

*Drexler:*   I would suggest that you collaborate. I am very interested in the idea of having an engineered structure on the tip of an STM- or AFM-positioning mechanism.[4] Do you see some way of using a bare STM tip, of the sort that are available today, to build molecular machines?

*Foster:*  Using a bare tip doesn't seem to be a good idea. They simply don't work very well. This is a huge amount of the unreliability of STMs and AFMs. I should also point out that the state of the art of AFMs is mixed and not the same as STMs. AFM work has much less resolution than STM work; currently you can't really see atoms with AFMs. That is especially true in its attractive mode; as you get nearer and feel the attractive van der Waals forces, and so on, the resolution deteriorates to about 20 Å. You can't move individual molecules that way.

*Drexler:*  What seem to be the physical limits in the repulsive regime?

*Foster:*  Mostly elasticity. When you push down on something at the atomic scale, it flexes. The load spreads out over the whole tip.

*Drexler:*  So better control over tip structure combined with better sensitivity to force might overcome that limitation and approach atomic resolution.

*Foster:*  Yes. Or you can work on a more conductive surface and use electrons for control, and with electrons you can control down to the last atom.

### Notes

1. See J. Rothman, "Polypeptide chain binding proteins: catalysts of protein folding and related processes in cells," *Cell* 59 (17 November 1989): 591–601.

2. C. Noren, S. Anthony-Cahill, M. Griffith, P. Schultz, "A general method for site-specific incorporation of unnatural amino acids into proteins," *Science* 244 (14 April 1989): 182–188.

3. N. Seeman, "Nucleic acid junctions and lattices," *J. Theor. Biol.* 99 (21 November 1982): 237–247.

4. See K. E. Drexler, J. Foster, "Synthetic tips," *Nature* 343 (5 February 1990): 600.

See also K. E. Drexler, "Molecular tip arrays for AFM imaging and nanofabrication," Paper delivered at the Fifth International Conference on Scanning Tunneling Microscopy/Spectroscopy and First International Conference on Nanometer Scale Science and Technology, Baltimore, Maryland, July 1990.

# III

## Perspectives

# 12

## Possible Medical Applications of Nanotechnology: Hints from the Field of Aging Research

Gregory M. Fahy

My task here is to comment on potential medical applications of nanotechnology. I do this not as an employee of the American Red Cross but as an interested observer of both nanotechnology and aging research.

### Medical Uses of Nanotechnology

A few general perspectives on possible medical applications of nanotechnology are needed at the outset.

First, I must in all candor admit that this is a potentially vast subject about which little can be said in detail at the moment. This chapter will therefore necessarily differ somewhat in character from the more technical chapters in this volume, which focus on more established and hence more concrete realms of science, engineering, and exploratory engineering. One reason for focusing specifically on aging phenomena is to be able to bring as much specificity and detail to the discussion as possible—despite the fundamental problems of technological forecasting.

Second, it is essential in contemplating this subject to ask what we would like nanotechnology to do for us. It is reasonable to expect that a molecular device that includes a mainframe computer, such as Drexler describes, would be complex enough to do sophisticated molecular repairs and still be small enough to fit inside a typical mammalian cell. What is not clear is just what need we would have for such devices.

Clearly, the goals of medical nanotechnology are the same, in general, as the goals of conventional medicine: the enhanced physical and psychological well-being of humans (and whatever they may become in the future). The latter aspect, psychological, is an important one. The desire

**Table 12.1**    Medical Goals and their relationship to nanotechnology

| Difficulty level | Specific medical problem | Possible molecular remedies | Level of required molecular technology |
|---|---|---|---|
| Level 1 | Cellular disease reversal | Molecular diagnostics and treatments | Advanced "pharmacology" and gene therapy |
| Level 2 | Health maintenance | Enhanced self-repair and immunity | Advanced "pharmacology" and gene therapy |
| Level 3 | Morpho-engineering | Genetic engineering, regeneration, tissue fabrication | Gene therapy and large-scale phenotypic reprogramming |
| Level 4 | Restoration of nonhomeostatic tissue | Anabolocytes, molecular computers, cell repair machines, catabolism suppression | Nanotechnology |
| Level 5 | Invulnerability and bio-enhancement | Redesign and refabrication of the living human body | Nanotechnology |

for psychological well-being will result in some medical goals quite different from those that people have had in the past. The main difference between present-day medicine and medical nanotechnology is that the goals of medical nanotechnology would be realized through molecular-level alterations of the human body that cannot be achieved efficiently without the availability of general purpose, thorough control of the structure of living and nonliving matter. This would include derivative, higher-level alterations as well, such as growing organs in vitro using molecular fabrication technology and then introducing them into the body.

Table 12.1 lists general medical goals, ranging from "simple" problems, such as curing all known diseases, to much more ambitious challenges.

Starting at a level even below level 1 in table 12.1, what can be done? Modest extensions of present-day molecular technology will likely be used to devise diagnostic equipment: molecules that can report any abnormalities found, whether in vitro or in vivo. It is only a short step further to convert some of these diagnostic items into specific treatment modalities.

If we consider most diseases to be the result of infection and mutation, then it is likely that all it will take to dispose of such simple present diseases is a sophisticated level of molecular biological technology considerably short of nanotechnology.

The next rung on the ladder is maintenance of a healthy state. The only major threat to health—after diseases have been conquered—is the aging process. The level of technology required may also fall short of nanotechnology; simply enhancing self-repair and successfully maintaining immunological function may solve the problem.

Beyond this, the psychological side of medical goals begins to become important. After you become a permanently healthy immortal, what other medical goals can you have? Well, people will still have their vanity to satisfy. They may wish to change their adult body size, to lengthen just their legs, to remodel the shape of their skulls, and so on. Beyond this, accident victims may require regeneration of crushed, burned, or severed body parts, and people disfigured by birth defects—before disease reversal precludes such defects—or people who are otherwise damaged, will require help. These goals represent major departures from what I would call a mature "entry-level" genetic engineering; they cannot be achieved by turning on or turning off a few existing genes or by synthesizing and inserting a few additional genes that simply lead to the production of ordinary types of catalytic proteins. Instead, actualizing goals of this nature will require both an understanding of how to program living tissue to create desired macroscopic structures (something the body already knows how to do) and a means of assuring that the process of overriding previous, natural, morphogenetic programs goes smoothly. This is hardly a trivial undertaking but one that can still rely on relatively ordinary biological mechanisms. Nanotechnology would likely be helpful for these tasks, but goals at this level can probably be achieved, although with some difficulty, without a mature nanotechnology.

What else might be medically relevant beyond this? "Fatal" traumatic accidents, attempted homicides, massive exposure to radiation, prolonged accidental systemic poisoning, and so on, could all lead to prolonged cardiac arrest and/or "irreversible" cellular damage, with or without a variety of other systemic traumas superimposed. This problem is qualitatively different from any considered so far due to our inability to rely on cellular homeostatic mechanisms to maintain "housekeeping" while structural repairs are carried out. Deterioration prior to treatment presents additional problems. No technology short of nanotechnology would seem to have any hope of successfully addressing these difficulties, and we must expect that even nanotechnology will fail in many cases.

The final goal in table 12.1 is to redesign and refabricate an existing, already living human body so as to make it far less vulnerable to injuries of all kinds (with, for example, subcutaneous "chain mail," "unbreakable" bones, more chemically resistant skin, and cells that do not swell when deprived of energy) as well as to endow it with capabilities no human body has ever possessed (such as super intelligence, super strength, and vision at unconventional wavelengths). One impetus that will certainly move us toward such goals in the long run is the desire to conquer space and other environments for which humans are presently poorly adapted. Although these medical goals may be viewed as outgrowths of level 3, they are much more open-ended and of a much deeper nature which places them firmly at level 5.

You may now appreciate the difficulty of presenting specific medical applications of nanotechnology at this time. However, what can be done, and what the rest of this chapter attempts, is to fill in the first couple rungs on the ladder in order to establish a baseline, and thereby perhaps prepare you to take the nanotechnological applications more seriously.

## Molecular Approaches to Disease

Consider the antibody. An antibody is an excellent sensor. It can latch onto any molecular structure outside of a range of structures that the immune system has been trained to accept as "normal." Very recently, since 1986, antibodies have been used in some very imaginative work. The term *abzymes* has been used to refer to antibodies that do not just find things but catalyze reactions once they find a particular something. Abzymes can be made in a variety of ways. One technique is to synthesize a molecule that looks like the transition state of a reaction that you want to catalyze and use it as an immunogen to elicit antibodies that bind to it. When such an antibody binds an appropriate substrate, it induces a change in the structure of the substrate toward the transition state and thus catalyzes the reaction. Molecules have been designed and created to elicit abzymes that can cleave certain peptide bonds that are not cleaved by existing enzymes, such as the bond between glycine and phenylalanine.[1] Abzymes have been made that will take a linear molecule and make it into a ring structure. In another application, reported recently in *Science News,* an abzyme was used to reverse thymine dimerization, which is a kind of DNA damage caused by UV light.[2] It is thus not hard to imagine abzymes roaming the body, not only "noticing" the presence of bacteria or viruses, but then starting to destroy them.

That is just one simple concept about how molecular machinery could be used to deal with existing diseases. More sophisticated approaches such as genetic therapies to replace defective genes and/or to cure cancer are being developed as I speak, and some are nearing clinical application already. Most of the diseases we have today are either the result of external invaders or defective genetic regulation. Hence, sooner or later relatively standard molecular tools should be able to deal with these problems.

## Aging

### General Features of Aging

The remainder of this chapter deals with the problems of aging (level 2 in table 12.1). The aging process has been postulated to arise in either of two ways. The first is through random damage, such as free radical damage or nonenzymatic glucosylation of sensitive sites on proteins. However, rather than random damage, physiological changes are usually found such as changes in neurotransmitter and hormonal receptor population sizes. Much that resembles random damage occurs as a consequence of these physiological changes. Therefore the lion's share of aging seems to be due to the second postulated process: genetically "programmed" senescence. In all likelihood, both processes play at least some role. Nevertheless, as the following discussion will suggest, "physiological" and relatively ordinary intervention strategies should be equal to the task of dealing with either type of problem. Nanotechnological computer-effector systems would certainly provide effective ways for dealing with random damage. But other techniques—based on the observation that cells are actually very good at repairing themselves—may provide satisfactory solutions. Incorporating a few additional repair systems may be all that is needed to handle any damage that results in aging.

Are there any precedents that would make us believe that it is possible to eliminate aging? The answer is affirmative. There are living trees that are several thousand years old that are perfectly healthy and show no signs of aging.[3] There appear to be nonaging animals as well, particularly sea anemones,[4] hydras, certain lobsters, and possibly certain fish. (Data are somewhat limited; it is hard for short-lived creatures such as humans to monitor individual free-living ocean-dwelling animals for several decades.)

### Molecular Details of Aging

To illustrate the nature of the problem of controlling aging, I will consider here five key molecules that are central to the aging process. All control major life-maintenance systems and all are virtually lost ($\sim 70\%$ declines) with age.

1. Coenzyme $Q_{10}$ is a key molecule in energy metabolism in every cell in the body except for red blood cells (which do not have mitochondria).
2. Dehydroepiandrosterone (DHEA) is the precursor for steroid hormones and inhibits age-related diseases.
3. Human growth hormone may control the immune system; manipulating this molecule could allow immunity to continue indefinitely instead of declining with age.
4. DNA polymerase alpha, which is primarily responsible for DNA synthesis and repair, undergoes bizarre changes with aging.
5. Elongation factor 1, which essentially determines what happens to protein synthesis as we get older. Since protein synthesis is required to produce the previous four molecules (and virtually *every* molecule in our cells), control of protein synthesis is particularly important.

If we could prevent the loss of these five molecules, we would be a long way toward controlling aging—even without nanotechnology. On

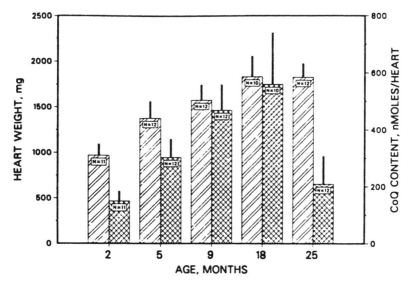

**Figure 12.1**
Heart weight and coenzyme $Q_{10}$ content versus age. (Reprinted with permission from R. Beyer et al., "Tissue coenzyme Q (ubiquinone) and protein concentrations over the life span of the laboratory rat," *Mech. Aging and Dev.*, 32 (1985): 267–281. Copyright 1985 Elsevier Scientific Publishers, Ltd. Ireland.)

the other hand, molecular engineers and aspiring nanotechnologists can, if desired, use these five pivotal molecules to establish concrete, specific design goals for gerolytic molecular engineering technology.

**Coenzyme $Q_{10}$**   Coenzyme $Q_{10}$ consists of a quinone moiety, which carries a free radical from cytochrome to cytochrome within the mitochondrial membrane, and a polyprenyl tail. Coenzyme $Q_{10}$ shuttles back and forth between the different elements of the electron transport chain to provide normal energy metabolism. Coenzyme $Q_{10}$ (or molecules in which the length of the polyprenyl tail has been modified to match the thickness of different types of mitochondrial membrane) is found in every aerobic cell on the planet. In fact, you could not live without coenzyme $Q_{10}$ in your body.

In this regard, it is interesting that just before the age at which most animals start to die from aging, cardiac coenzyme $Q_{10}$ levels suddenly drop by about 70% (figure 12.1).[5] We see similar findings in human

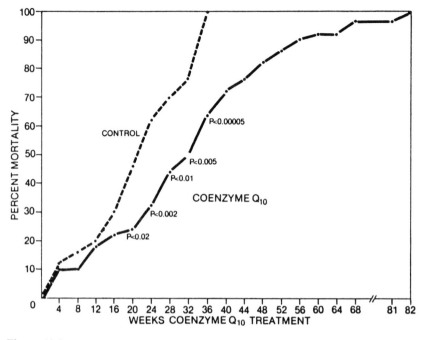

**Figure 12.2**
Effect of coenzyme $Q_{10}$ on life expectancy of female mice. (Reprinted with permission from E. Bliznakov, "Coenzyme Q, the immune system, and aging," *Biomedical and Clinical Aspects of Coenzyme Q*, vol. 3, ed. K. Folkers, Y. Yamamura. Copyright 1981 Elsevier North Holland.)

beings.[6] Coenzyme $Q_{10}$ deficiencies are worst in people who go on to die of heart disease. If you give an animal coenzyme $Q_{10}$ shortly before it enters the phase of its life cycle where mortality accelerates in the population, you can extend the survival time of the animal considerably. This has been achieved in female mice; coenzyme $Q_{10}$ was given by injection at a level that is roughly equivalent to one-third the customary human pharmaceutical dose of 30 mg/day used in Japan (figure 12.2).[7]

These results suggest that a molecular machine capable of interacting with the appropriate DNA targets could prevent the responsible genes from turning off coenzyme $Q_{10}$ synthesis as part of the aging process. Alternatively, it might be possible (and a lot simpler) to slow down the turnover (i.e., the catabolism) of coenzyme $Q_{10}$. Machines that carry out both tasks already exist and are constantly regulating the synthesis and breakdown of coenzyme $Q_{10}$ and other molecules. What we need are techniques for modulating the existing relevant molecular machines and ensuring that they continue to function properly.

**Dehydroepiandrosterone (DHEA)**    DHEA is the major adrenal steroid in humans. For a long time, nobody had any idea what it did. It does not do the sorts of things that adrenal steroids usually do. Its basic role is to reduce the efficiency with which food calories are converted into fat. As we age, DHEA levels drop drastically, and as a result, we tend to get fatter and lose muscle tissue. DHEA also has an antiproliferative effect; it tends to inhibit atherosclerosis and cancer. Levels of DHEA circulating in the bloodstream are inversely proportional to cardiovascular mortality and, in women, to the risk of breast cancer. This has inspired the efforts of several pharmaceutical companies to produce derivatives of DHEA that can be patented.

However, because we would prefer not to take a drug, future molecular tinkering that could maintain the synthesis of DHEA at youthful levels, or slow down its catabolism, as with coenzyme $Q_{10}$, would be preferred.

**Growth Hormone**    The immune system is fundamental to life maintenance. The thymus gland, located under the breastbone, is genetically programmed to respond to growth hormone. After puberty, as growth hormone levels drop off, the thymus gland begins to atrophy. The hormones that the thymus gland produces, which provide education for the immune system, also decline, and the diseases of aging increase coordinately.

It is possible to regenerate the thymus gland by giving aging animals implants of pituitary tumor cells that give off large amounts of growth hormone.[8] A 24-month-old rat has a greatly deteriorated thymus gland compared to a 3-month-old rat, and only about one percent of the immune function of the younger rat. Implantation of the pituitary tumor into the older rat causes a partial regeneration of the thymus gland and a ten-fold increase in immune function. If the same experiment is done with slightly younger (18-month-old) rats, the result is total thymus gland regeneration and total restoration of immune system competence. Follow-up studies have shown that similar results can be obtained in dogs by giving growth hormone directly. In other experiments, monthly transplants of neonatal thymus glands into aging recipients resulted in a 46 percent increase in mean life span. Obviously, we would like to prevent the 70 percent decline in growth hormone that occurs in humans as a result of the aging process.

**DNA polymerase alpha (DPA)**    DNA polymerase alpha is required for most DNA repair and DNA synthesis. Like the processes of energy metabolism and immunity already discussed, DNA repair and synthesis are obviously fundamental life processes without which we would die quite quickly. It is therefore bizarre but true that we each have a gene for a defective version of DPA that switches on as we get older. The senescent or "old-age" form of the enzyme is called A1 DPA. It binds to DNA much more weakly than the youthful or A2 form of DPA and, consequently, it synthesizes and repairs DNA ten times more slowly than A2 DPA. In addition, it makes an alarming number of mistakes. As we age, A1 DPA replaces more and more of our A2 DPA, until we have almost no A2 DPA at around the age of 65.

Equally amazing, it is possible to convert the senescent A1 DPA into a form that closely resembles A2 DPA by treating it with inositol-1,4-bisphosphate, a sugar phosphate produced by the hormonal stimulation of cells.[9] The natural transition from A2 DPA to A1 DPA production is a programmed aging phenomenon that is measurably more advanced in 40-year-olds compared to 30-year-olds. We should be able to do something to prevent this process, presumably by altering the control regions for the A1 and A2 DPA genes or by adequately maintaining particular hormone levels such as growth hormone.

Although it has been suggested that shutting down cell division is a protective device for minimizing cancer, thereby making the A2 to A1 transition good for us, misincorporation of DNA bases can only promote cancer. Furthermore, cancer may not be a problem regardless of DNA damage if we can maintain the immune systems of 18-year-olds, which probably eradicate cancer all the time without our awareness.

**Elongation Factor One (EF-1)**    George Webster has shown that there is a massive drop in protein synthesis in all tissues just before the age at which animal death rates accelerate (figure 12.3).[10] By studying (directly

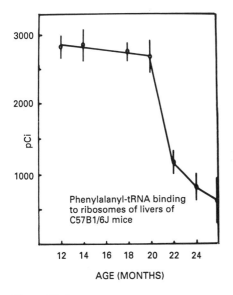

**Figure 12.3**
Protein synthesis versus age. (Reprinted with permission from C. Blazejowski,
G. Webster, "Decreased rates of protein synthesis by cell-free preparations
from different organs of aging mice," *Mech. Ageing Dev.,* 21 (1983): 345–
356. Copyright 1983 Elsevier Scientific Publishers, Ltd. Ireland.)

or indirectly) all of the more than 50 molecular machines involved in
the complex biosynthesis of proteins, he identified one molecule that
was responsible for about half of this drop in protein synthesis in most
organisms (rats, mice, and flies).[11]

Protein synthesis from available mRNA consists of three component
processes: initiation, elongation, and termination. The portion of the
protein synthesis process that decreases precipitously with age is elongation. Elongation in turn consists of three steps: (1) the amino acyl
tRNA binds to the A site of the ribosome; (2) a new peptide bond forms
that links the new amino acid with the nascent peptide chain, thereby
transferring the nascent peptide to the A site and releasing the cleaved
tRNA (the transpeptidase reaction), and (3) the peptide with the added
amino acid moves from the A site to the P site so that the next amino
acyl tRNA can bind to the A site. With aging, the gene for elongation
factor one (EF-1) is selectively shut off, effectively blocking the first step
of the elongation process (figure 12.4).

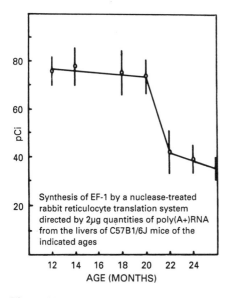

Synthesis of EF-1 by a nuclease-treated rabbit reticulocyte translation system directed by 2µg quantities of poly(A+)RNA from the livers of C57B1/6J mice of the indicated ages

AGE (MONTHS)

**Figure 12.4**
Decrease in EF-1 mRNA with age. (Reprinted with permission from G. Webster, "Protein synthesis in aging organisms," *Molecular Biology of Aging: Gene Stability and Gene Expression,* ed. R. Sohal, et al. 1985. Copyright 1985 Raven Press, Ltd.)

EF-1 is a major constituent of the cell; it comprises between 3 and 11 percent of the total protein of a nonsenescent cell. We would like to be able to prevent the specific repression of this protein with age. The decreased protein synthesis caused by the repression of EF-1 may cause a drop in RNA synthesis since RNA polymerase is a protein. Indeed, global drops in RNA synthesis are usually seen with aging.

Webster has shown that administration of centrophenoxine can fully restore the massive drop in RNA synthesis that takes place with aging (figure 12.5).[12] In preliminary, unpublished experiments, he has also found evidence that this drug can specifically reactivate transcription of EF-1 genes. Although Webster's results were obtained using a chromatin assay system in a test tube, centrophenoxine has also been shown to reactivate RNA and protein synthesis in vivo, in mammals, and to extend mammalian life span. However, its effectiveness is limited by its rapid hydrolysis in the body (see also figure 12.5). The most elementary form of interventive molecular engineering imaginable would be to design a

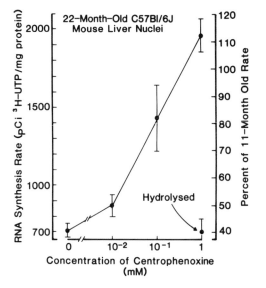

**Figure 12.5**
Restoration of RNA synthesis by centrophenoxine.

stable version of centrophenoxine that lacks a labile ester linkage yet still possesses the biological effects of centrophenoxine. This single trivial accomplishment could lead to major gains against aging.

## Conclusion

There is good reason to believe that future medicine can look at least as amazing as Drexler has projected based on his ideas about nanotechnology. We can hope that eventually molecular nanotechnology will assume the role of the fanciful molecular repairman shown in figure 12.6. This molecular engineer looks to me like the ultimate Santa Claus, looking at every molecule to determine if it has been naughty or nice— and repairing it if necessary. Even without this ultimate technology, I think we can do a great deal; with it there should be virtually no limit to what we can do.

**Figure 12.6**
Drexler's Demon (with apologies to Maxwell).

## Notes

1. B. Iverson, R. Lerner, "Sequence-specific peptide cleavage catalyzed by an antibody," *Science* 243 (3 March 1989): 1184–1188.

2. I. Amato, "Teaching antibodies new tricks: Antibodies that act like enzymes are filling chemists' heads with new visions," *Science News* 136 (1989): 152–153, 155.

3. G. Touzet, "Perennial plants," *Interdisciplinary Topics Geront.* 21 (1985): 263–283.

4. A. Comfort, *The Process of Aging* (New York: New American Library of World Literature, 1964).

5. R. Beyer, B. Burnett, K. Cartwright, D. Edington, M. Falzon, K. Kreitman, T. Kuhn, B. Ramp, S. Rhee, M. Rosenwasser, M. Stein, L. An, "Tissue coenzyme Q (ubiquinone) and protein concentrations over the life span of the laboratory rat," *Mech. Ageing Dev.* 32 (1985): 267–281.

6. K. Folkers, S. Vadhanavikit, A. Mortensen, "Biochemical rationale and myocardial tissue data on the effective therapy of cardiomyopathy with coenzyme $Q_{10}$" *Proc. Natl. Acad. Sci.* 62 (1985): 901–904.

7. E. Bliznakov, "Coenzyme Q, the immune system and aging," *Biomedical and Clinical Aspects of Coenzyme Q*, vol. 3, ed. K. Folkers, Y. Yamamura, (New York: Elsevier North Holland, 1981), 311–323.

8. K. Kelley, S. Brief, H. Westly, J. Novakofski, P. Bechtel, J. Simon, E. Walker, "$GH_3$ pituitary adenoma cells can reverse thymic aging in rats," *Proc. Natl. Acad. Sci.* 83 (1986): 5663–5667.

9. V. Sylvia, G. Curtin, J. Norman, J. Stec, D. Busbee, "Activation of a low specific activity form of DNA polymerase alpha by inositol-1,4-bisphosphate," *Cell* 54 (1988): 651–658.

10. C. Blazejowski, G. Webster, "Decreased rates of protein synthesis by cell-free preparations from different organs of aging mice," *Mech. Ageing Dev.* 21 (1983): 345–356.

11. G. Webster, "Protein synthesis in aging organisms," *Molecular Biology of Aging, Gene Stability, and Gene Expression,* ed. R. Sohal, L. Birnbaum, R. Cutler (New York: Raven Press, 1985), 263–289.

12. G. Webster, "Protein synthesis," *Drosophila as a Model Organism for Ageing Studies,* ed. F. Lints, M. Soliman (London: Blackie, 1988), 119–128.

13. R. Sohal, R. Allen, "Oxidative stress as a causal factor in differentiation and aging: A unifying hypothesis," *Exp. Geront.* 25 (1990): 499–522
R. Allen, "Oxygen-reactive species and antioxidant responses during development: The metabolic paradox of cellular differentiation," *Proc. Soc. Exp. Biol. Med.* 196 (1991): 117–129.

14. G. Lenaz, ed., *Biochemistry, Bioenergetics and Clinical Aspects of Ubiquinone* (London: Taylor & Francis, 1990).

15. M. Kalimi, W. Regelson, eds., *The Biologic Role of Dehydroepiandrosterone (DHEA),* (New York: DeGruyter, 1990).

16. D. Rudman, A. Feller, H. Nagraj, G. Gergans, P. Lalitha, A. Goldberg, R. Schlenker, L. Cohn, I. Rudman, D. Mattson, "Effects of human growth hormone in men over 60 years old," *New England J. Med.* 323 (1990): 1–6.

17. M. West, "Reversing cellular aging in connective tissues," *Life Extension Rep.* 10 (1990): 73–75.

18. J. Shepard, U. Walldorf, P. Hug, W. Gehring, "Fruit flies with additional expression of the elongation factor EF-1 alpha live longer," *Proc. Natl. Acad. Sci. USA* 86 (1989): 7520–7521.

19. G. M. Fahy, "Short-term and long-term possibilities for interventive gerontology," *Mount Sinai J. Med.* 58 (1991): 328–340.

## Discussion

*Drexler:* The bulk of your remarks relate to pharmaceutical chemistry and design. Would you care to comment on the relationship between molecular engineering in the pharmaceutical industry and molecular engineering in general?

*Fahy:* I think that it is all a continuum. If you have capability for doing one kind of molecular engineering, it leads to capability for other kinds as well. Pharmaceutical companies are a nice place to start with this sort of approach because they have everything to gain by designing new ways of manipulating biological systems. The sort of full-blown nanotechnology that we expect in the future might develop incrementally from a large number of small steps on behalf of such companies.

*Schwartz:* I have always felt that I would have the option to live to be 150. You may be pointing in the direction that will prove me right.

## Postscript

In the two years since the First Foresight Conference on Nanotechnology, several events have taken place that further strengthen and reinforce my specific and general points about molecular medicine, aging, and nanotechnology.

Molecular approaches to disease have advanced considerably. The first human clinical trials of gene therapy have begun and are yielding exciting preliminary results. Antisense mRNA technology has been successful at greatly decelerating the in vitro cellular aging clock for human cells (more than doubling the number of cell divisions that can occur before cell senescence occurs). This result is now being developed into a commercial product. An important new molecular handle on cancer has come from the discovery of the pervasive role of mutations in the p53 cancer suppressor gene as prerequisites for malignancy. If antisense mRNA can block cellular aging, can "sense" p53 mRNA replace the function of mutated p53 and lead to the arrest of a broad spectrum of cancers without side effects? Antibodies are being used to block the rejection of transplanted tissues and to provide a means of decorating malignant cells with enzymes that convert harmless prodrugs into potent chemotherapeutic agents directly on the surfaces of these malignant cells. More attention is being placed on the possible diagnostic use of molecular technologies such as the polymerase chain reaction for such applications as detecting HIV infection.

With respect to our global understanding of aging, at least two very noteworthy developments have supported the notions I described. First, the appearance of Caleb Finch's truly epic book, *Longevity, Senescence, and the Genome* (Chicago: Chicago University Press. 1990), a treasure trove of aging information, provides strong additional evidence for a predominantly genetic basis of aging. Second, a most unexpected synthesis has been presented by Sohal and Allen.[13] They posit that the progression of development, and hence also the progression of aging, is regulated by a redox-stat that gradually renders cytosol more oxidizing with time. Their synthesis, if correct, would show how past evidence for wear and tear is actually evidence for genetic control of aging.

The specific examples of intervention into aging I presented have been confirmed and extended—in some cases with startling speed. Coenzyme

$Q_{10}$ has received significant new recognition from the medical community (and new ire from the FDA due to its use without FDA regulation). Two new technical books on its biomedical and clinical applications have appeared along with one new popular book.[14] An entire book has also been published on DHEA.[15] Certainly the most noteworthy development, however, has been the much-publicized explosion of studies and attention directed to the administration of human growth hormone (HGH) to elderly individuals in an attempt to reverse many aspects of human aging, not just immunological functions. Rudman's claim that HGH appeared to reverse 20 years' worth of aging according to some measures was considered too flamboyant by many but appears to have been accurate (he examined muscle mass, fat mass, skin thickness, bone density, and the size of internal visceral organs).[16]

My inference based on the DNA polymerase alpha story that it should be possible to prevent replicative senescence and compromised DNA repair appears to have been confirmed beyond all reasonable expectations by Mike West's findings. West claims to have totally conquered the cellular aging clock by using just two naturally occurring molecular triggers that can restore the youthful phenotype to any one of four entirely different types of senescent cell (endothelial cells, astrocytes, chondrocytes, and fibroblasts).[17] Finally, the possible significance of elongation factor one has been delightfully demonstrated by a paper of Shepard et al. in which flies made transgenic for EF-1 so as to avoid or postpone its down regulation with age were shown to have mean life spans equivalent to the maximum life spans of control flies.[18] However, this experiment is not yet definitive and awaits rigorous verification.

It appears that progress in both molecular technology in general and in interventive gerontology in particular is heading very much in the general directions I projected, but at considerably greater speed than I had anticipated. If this conclusion sounds familiar, so be it. I think we can expect to continue to be surprised by the fast pace of progress for some time to come.

It is my pleasure to acknowledge that it was interest in the proceedings of the First Foresight Conference on Nanotechnology that led to an invitation to greatly expand my proceedings paper into a much fuller and much more completely documented work.[19]

# 13

# The Future of Computation

Bill Joy

Although I am not a scientist, I have worked in the area of scientific computation for quite some time. In 1972 and 1973 I worked with early prototypes of the first Cray computer, writing very large programs for scientific problems. After that, I went to Berkeley and did some work on UNIX. In 1982 we started a computer company, Sun Microsystems, which is now the leading vendor of scientific and technical workstations. So I have had an abiding interest in providing high-performance computers for people who want to do scientific and engineering research.

## In 1982, a Gloomy Prospect

When we started Sun, we were disturbed about what we saw as a marketplace trend. The hot company at that time was Atari, which made computers with incredible graphics compared to what the average scientist had. The way the game and consumer market was going threatened to create a situation where kids at home would have better graphics than scientists doing research. I did not find that acceptable, but if you look at the economics, and the volume that Atari was shipping, you could easily see why this was happening.

At the same time, the large companies in the computer industry, IBM and DEC, were abandoning the technical marketplace to concentrate on the commercial marketplace. Hence the ten-year gap between the dominance of vector computation and the introduction of a vector accelerator for the VAX.

The continuing trend is that not many people care very much about making high-performance scientific computers; they are usually a side effect of developing computers for something else. Sadly, it takes a lot of investment to make a good scientific computer. Historically, I think the best scientific computers have been built by scientists themselves. People doing science cobble together microprocessors and in many cases make machines that would put a lot of computer research labs to shame. Scientists are very creative, and the computers are well engineered, reflecting that the people building the machines have a profound interest in the problems they are trying to solve. The computer is secondary to the problem; so it gets built well. A marketing department cannot hide poor design; if the computer doesn't work you don't get your research results.

What we've seen over the last 30 years is a distressing trend in that computers haven't really gotten much faster. They have only gotten smaller and cheaper. In fact, it probably wasn't long after 1960 that you could get a machine that was about a one MIPS (million instructions per second) computer. If you had a big enough room, you could also get a megabyte of memory and a disk drive. This is fundamentally the environment that dominates most interactive software today. There is a disk drive spinning at some rate and every 30th or 60th of a second you get some data from it. You have about a million memory locations, and you can execute about a million instructions a second. These are the fundamental constants that affect the way people think when they write an interactive application. It is a real constraint on what people can do. It has created what Neil Lincoln, a designer for many years at CDC and then at ETA, called "the VAX generation," people who think that if they have one VAX computer they can do anything.

That thinking sets the limits of the vision for these people, and this limit is clearly not good enough. With a million instructions per second, and, given that when a person is typing or doing some other interactive activity with a computer, they are willing to wait only one tenth of a second for a response to a trivial command, you can execute about 100,000 machine instructions on a one MIPS processor before you have to give the user a response. This is fine for echoing a character on a WYSIWYG word processor, or for doing something simple on a spreadsheet. But you cannot do very much scientific computation in 100,000 instructions. The complexity of the alternate reality creation that you can create in the inner loop of a machine of this type is very limited.

When we started Sun, I had a particular goal: to drive the industry past the one MIPS, one megabyte, one disk drive bottleneck. This bottleneck did not matter for the people with word processors and spreadsheets because they could just write their applications in assembly language and cram them inside these limits. But if you want to do science, and you expect to do this by giving lots of bright people computers, you need to drive technology forward. Driving technology forward requires a goal that will stimulate the necessary investment within the industry, over a sustained period of time, in the technologies needed for the desired performance increase.

So we set a goal in the early 1980s to double the performance of a desktop microprocessor every year into the foreseeable future. We had a formula for the number of MIPS we wanted on a desktop: $2^{(year - 1984)}$.

This seemed crazy when we introduced it in 1983, but it has been pretty much the case that people have at least announced machines (they could not quite ship them) that have stayed on this curve. It has challenged people to drive price/performance forward. One of the reasons Intel—probably prematurely—announced the 486 chip was because of the pressure from people building reduced instruction set computing (RISC) microprocessors. These are the particular fashion that have been driving performance forward lately. In general, this performance increase has made much more powerful machines available. Where this matters the most, of course, is in science and engineering.

From the constant performance available in the 1960s, 1970s, and the early part of the 1980s, for the average interactive program of about a million instructions per second, we will, in the 1990s move to a typical interactive performance of a hundred million instructions per second (100 MIPS). This increase makes a huge difference in what we can do interactively. We can sift through a hundred times as much data in an interactive loop. With a good algorithm and a good data structure, the difference between 100,000 and 10 million instructions is enormous. Changing our transportation speed from 50 mph to 500 mph to 5000 mph has lead to an enormous transformation of our perception of space. A car allows you to drive from coast to coast in a few days; a jet plane gets you there fast enough so that you can essentially commute from California to Washington or New York—if you're willing to stay up for a very long day. This change brought about an enormous economic transformation.

In the same way, the existence of an "average machine" with a hundred times the interactive performance of our current machines will dramatically change what we conceive of doing interactively. Hundred MIPS machines and the kind of interactive graphics hardware now available on even inexpensive workstations represent the success of the vision of getting past the one MIPS bottleneck of the 1980s.

## Large-Scale Computing

In the early 1980s we were also looking for a similar goal for large-scale simulation engines and large-scale database engines to allow us to do practical things with large databases—such as mapping the human genome. Larry Smarr, who started the supercomputer center at the University of Illinois, and others involved in supercomputing were seeking the same goal. Unfortunately, I do not think this goal has been achieved. We seem, in fact, to be going backward. We've seen the demise of ETA systems; Cray is not doing well; one of the supercomputing centers is folding. The money is just not being invested to create the devices that can attack the scale of problems that are important for doing this sort of interactive, computational science.

Several years ago I tried to capture a simple-to-understand scale for measuring the performance of large-scale computers. People get excited about the increases in MIPS, but in fact a simple doubling of the MIPS power of a machine doesn't make that much difference in the kind of scientific problems that you can solve. So I created a scale that I've named the Wilson scale, after Ken Wilson, the Nobel laureate who said that he wanted a machine that ran at $10^{14}$ MIPS. This is such a large number that it is difficult to understand what it means. So I said that the Wilson scale should be $W = \log_{100}MIPS$, which puts the VAX at roughly $W = 0$ (which expressed my personal feelings about how much of the science I wanted to do could be done on a VAX). A Cray is about $W = 1$, which means that you can do one unit of Wilsonian computing on a Cray. You can work one- and two-dimensional problems on a Cray, but it is not really adequate for most problems in three dimensions. On this scale, what Ken wanted is at magnitude 7. His goal was to get what

I characterized as a Wilson 7 machine by the year 2001. He suggested that it was certainly a problem worthy of devoting the resources of the nation to, and, if necessary, spending $10 billion to get this kind of supercomputing resource. Essentially, such a machine would be a very powerful computational telescope or microscope for science.

On this scale from 0 to 7, what would be a 2? What would be 100 times as fast as a Cray? Not many things that are even 100 times as fast as a Cray 1. Machines like the Caltech Hypercube—these two- or three-dimensional stacks of microprocessors—if you consider the most powerful ones proposed, are about a Wilson 2. But it's a long way from 2 to 7. In 1986 I discovered Drexler's book, *Engines of Creation,* and, based on "back of the envelope" calculations, I realized that a mechanical nanocomputer capable of being a Wilson 7 device would measure about an inch on a side.

Why choose a factor of 100 to mark one step to the next? If you add a factor of 100 in computing power, you can generally do a one-step larger problem. If you have a two-dimensional problem, you can add a grid in three dimensions and step into the third dimension in 100 points. And that is a reasonable size grid. You can actually make some sense out of the additional dimension. Also, if you have another dimension, which is like a free variable, and you do not quite know what it is, with a factor of 100 increase, you can simulate a range of values and use search technique to find an optimum solution.

The time is past to have a reasonable chance of building a Wilson 7 machine by the year 2001 because we simply haven't started yet. A while ago the newspapers reported that the security agencies were going to build their own parallel supercomputers because they didn't expect to get them from the commercial sector. I think that the same thing might be true for scientific computing. Without large-scale investment by the government or by the commercial sector, it may have to be done by the physicists and the chemists, as it was in the 1950s.

## Speeding Up Computation

While I was preparing this chapter, Eric Drexler asked me to consider what I would do with a trillion processors. That actually scares me; if you ask me what I could do with a trillion processors, my first response is that I'd have a terrible time programming the machine. I've always looked at parallelism as a problem, not a solution. If possible, I'd rather take Thoreau's advice and "simplify, simplify, simplify."

There are many ways to get the answer to a computational problem faster. The best way is to get a better algorithm. This approach is effective independent of any other change. In the history of computing, some major problems have been speeded up more by improved algorithms than by improvements in technology. This will certainly be the case in the future.

You can also speed things up by cranking up the clock rate. Robert Birge describes a molecule responding to a photon in 500 femtoseconds (see chapter 7, "Molecular Electronics"). My sense is that it is not going to get much faster than that. Maybe you can get really clever with superconductivity circuits, but you are probably within a factor of 50 or 100 of the ultimate limit.

Another approach is to build a computing device with a new architecture. This is what the RISC movement has been about. Instead of making the clock rate faster, make the machine do more work per cycle. This is a nice technique because it's orthogonal to efforts in the semiconductor industry.

Finally, you can use parallelism. Of all of these methods, parallelism is the hardest. It's unfortunate, but it's true. It seems to be especially the case because each application of parallelism is a new problem. Some nice work has been done by Greg Fox's group at the California Institute of Technology looking at many problems on the machines they build there to understand, for each class of problem, how you can put it on a parallel machine.

But in order to make significant progress, I believe we need to focus a lot of bright people on one architecture instead of isolating small numbers of computer scientists on diverse parallel architectures. Without this focus there is no real benefit of one group's work upon the work of others because the machines are so different. If we had a national initiative to build one giant supercomputer, and if we forced people to decide which architecture would be used for the machine, then a lot of smart people could write algorithms that would work on the machine. What we have today is many high-performance architectures and no single body of algorithms, no reusable code, no standard that has emerged. The theme in the 1980s for much of the rest of computing has been the emergence of standards so that people could share their code. We don't have that for parallel computing. So, if we want a large-scale simulation capability, we need to have one large machine that everyone can focus on as a national resource.

What kind of compression could you have with a trillion processors? This is a difficult question. Consider it in terms of a VAX. If I had a machine that was 100,000 times faster than a VAX, I could do what I would do in a day on a VAX in less than a second because there are 86,400 seconds in a day. A Wilson 3 machine—a million times faster than a VAX—could reduce a VAX day to 0.1 seconds, the human response time. A VAX, a PC, a Macintosh, or an early Sun, are all about the same in performance. Anything you could compute in a day with one of those machines, you could compute between keystrokes with a Wilson 3 machine.

What about a Wilson 6? That's basically a VAX-century per frame (sixtieth of a second). You could make a movie such that each frame of the movie took 100 years to simulate on a VAX. It may seem like this is an absolutely ridiculous amount of computing power, but the kinds of animation done at PIXAR is in fact already nearly that much general-purpose computing per frame. (Each 70mm animation frame has 2400 × 3600 pixels.) If they had that much computing power, they could make a database of the characters, they could animate the personalities and bring them to "life." Currently, they use a lot of "hacks" to make the animations look right and reduce their need for computing, but this is roughly the amount of general-purpose calculation that they are using today. Since they obviously cannot afford to do animation today with a 100 VAX-centuries per second of frame time, they've built arrays of special-purpose multipliers and adders. It works very well for them because they get very effective use of that computer hardware, but we can't afford to build a special purpose machine for every problem.

Nanotechnology is exciting not because you could build a machine with a trillion processors but because you don't have to build that machine. You can create a molecular assembler and build a machine that is suitable for solving your particular problem. It seems far easier to invent a machine with the right shape to attack a problem than to figure out how to program an immense general-purpose machine. A second stage beyond building one machine so that you could write the algorithms only once would be to develop the technology so that you could build large-scale computers on a per problem basis. You wouldn't compile a program, you'd compile a computer. This is reminiscent of the "moties" in Larry Niven and Jerry Pournel's novel, *The Mote in God's Eye,* who must have had nanotechnology because they were continuously reshaping their spaceship to order. I have no difficulty conceiving of building a supercomputer simply for one run of a problem.

### Political Problems and Social Consequences

The problem with all of this is as much political as anything else. In order to build a big supercomputer that everyone can use, or to get the investment and infrastructure to construct even cellular automata—a machine that could put out any sort of cellular automata on a large scale—we need a large amount of committed capital. There is a lot of money going into RAM, certain types of microprocessors, television sets, telephones, and so on. But this doesn't directly attack the problem. We need to figure out how to develop political pressure for this research in the same way that we put political pressure on the computing industry, which led to open systems and standards and propelled us to 100 MIPS. It's not obvious to me how to do this.

It is conceivable that, in the next 20 years, we could carry around machines with $10^9$ processors. If you imagine each processor as $10^6$ transistors, that's $10^{15}$ bits. If a book is a megabit, that's $10^9$ books. Perhaps the most exciting thing for me is that I could have all of those books in a device that would be accessible to me and that I could query in any way I wanted. My favorite image of computers is the Apple vision: computers that are one-to-one with the user. One of the great things about nanotechnology is that we could use it to store and replicate this amount of information inexpensively. If we can get around the intellectual property problems by massive political pressure, we could have a machine the size of a sugar cube with all this information in it.

More important than the fact that it would be fast, cheap, and reliable would be its incredibly low demand for power. The real problem with computing devices today is that they tend to melt. Cray spends all of its time solving cooling problems. The amazing thing about molecular technologies, much like the earlier Josephson junction technology, is the much lower power requirements. Ultimately only low-power technologies will allow us to do vast amounts of computing in a small space.

My vision of getting past where we have been stuck for 30 years in computing is first to get to 100 MIPS on the desktop with great graphics—what we call super workstations. We came very close to achieving this in the 1980s. The next step is to get one $10 billion supercomputer built so that we can all play "what if" on some hard problems involving enormous amounts of data. We can use money to create a time warp! It is not very efficient to do things before they are really possible, but large enough amounts of money can make it happen (and $10 billion is not a large amount of money on the national scale, particularly if it were made politically attractive). We've been talking about a supercomputer research lab for a few years, but it hasn't gone anywhere yet, unfortunately. The next step after that is to build machines customized for each problem and to give people wonderful personal computers that fit in their pockets.

## Conclusions

Nanotechnology has the potential to be a very powerful force. I believe in this notion of future history, that what we do affects the distribution of possible futures. The great thing about nanotechnology is that you can imagine it being used to solve a lot of problems and to increase the probability of very good possible futures. The dangerous thing is that it pushes the distribution to be bimodal and introduces a lot of scary futures as well. Tomorrow is going to be very interesting. We have to decide how we are going to deal with this. It isn't just necessarily better; it's better and worse with some probability, and it's going to go one way or the other. We have to think about this. We can't simply do our science and not worry about the ethical issues.

When Drexler's book came out, I gave away several hundred copies; I saw it as an antidote to hackers never going home. Reading this book you suddenly realize that a homework assignment like, "implement a vampire" is not an unthinkable assignment anymore. You can start to see mechanistic explanations for things you've read about only in science fiction. Maybe people will relax a little and go and enjoy their lives, because we can see that we are going to get to those places eventually—if the planet survives. The exact rate at which we get there isn't important; we can enjoy things as we go along. That's the nicest thing about the concept of nanotechnology.

**Discussion**

*Drexler:*  You're right; a Wilson 7 machine would be about a one-inch cube with a kilowatt of power dissipation.

*Audience:*  What about the time frame for your projected parallel supercomputer program?

*Joy:*  We thought it could be a 15-year program starting in 1985. The cost is mainly the expense of assembling enough modules. It would be cheaper if you waited longer, but at some point you have to say that the value of having the machine is worth building it—even though building it later will always be much cheaper. It's not really that large an expenditure, about the same as the superconducting supercollider (SSC). There is a fair argument to be made that if you had a big enough computer, you wouldn't need the SSC. But that depends on how smart your heuristic search is. No one has ever had a computer that fast that could generate random hypotheses and test them. Brute force doesn't always win of course: Kasparov can still beat the best chess computer. But linear search is still a great technique—put everything in memory and run a search. There would be no I/O (input-output) problem because everything would be in memory.

*Audience:*  Could you compare this proposal with the Strategic Defense Initiative (SDI) test bed?

*Joy:*  I don't think the SDI test bed will ever be built. This was a plan to buy large numbers of Cray computers to try to track thousands of warheads. I doubt that they could have written the code to do that anyway.

*Audience:*  Why do we really need a national project to build a large computer? Where will the money to program it come from?

*Joy:*  You need to have a large machine in existence as a driving force to get people to write parallel software and to think about solving big problems. It's like the moon shot. It gets people focused on science; it gets them excited. It allows people to ask big questions. I honestly believe that a lot of the big questions will be solved by somebody really clever working on a small computer—if everybody had a computer powerful enough to do some basic science. If you give 100,000 kids or 100,000 researchers 100 MIPS graphic supercomputers, that would make as much difference as the large $10 billion supercomputer. But I believe that the first thing is going to happen anyway, so if you're looking for something to make an incremental difference, a big machine is the way to go. Money to program it will exist as part of the money to buy it. This is how the NSF is doing it with the supercomputer centers. Having the machine as a target, as a reason to write software and as a reason to automate science, will make a huge difference.

*Audience:*   Why should a politician spend $10 billion on this project instead of on something else?

*Joy:*   You have to scare people into doing it since we already have more than a $200 billion deficit. It has to be something like a computer race with Japan or it won't happen. It really isn't a zero-sum game; it's a question of who pays and when. If we don't do something like this, we'll end up paying because we won't be productive enough to pay off the national debt in any reasonable way. Our science will fall behind like our manufacturing did because we aren't investing enough in science. This project will force the development of science.

Science in the twenty-first century will be computational science as much as anything else. Fairly soon, a Nobel Prize will be won by someone whose results are computationally derived. Truly difficult problems require computing, and the kinds of problems that can be solved will be limited by the best computers available—and they are not very good right now. The research and development necessary to improve our computing tools can't be provided by the private sector. For example, Cray only ships one billion dollars worth of supercomputers a year. They can't spend several billion dollars on research and development. And the best supercomputer designers are out of work right now because the industry is not doing well.

*Audience:*   Will better networking of present computers make a super-large computer unnecessary?

*Joy:*   If it works, it's a help. The companies implementing networks using semiconductors have done badly of late. We don't have enough money invested in high-performance networking or computing. It's a sad situation in that we are waiting and hoping for the components for optical computers to trickle down from telephone development. Some companies are doing some relevant work in molecular technologies and other advanced computing technologies, but whether they pursue these vigorously depends on their business plans.

You only hit what you aim at. We need a program to build a powerful computer. I don't have confidence in a strategy that depends on getting the computer as a side effect of something else.

*Audience:*   Other than creating fears of a technology race, what could be used to justify the need for a very large computer?

*Joy:*   We need the machine to solve problems that we have on a planetary scale. The question is to find some motive force to drive the political process. Studying the environmental crisis is psychologically the most likely way to get it through Washington. For example, we have very little idea of what's going on in the atmosphere. Large amounts of data from satellites is going unread because no one has a computer large enough to put the data on line or to analyze it, so we can't build a model to simulate the atmosphere. We aren't even training people who could ask the right questions of the data if we did have a computer capable of analyzing the data. It's a question of infrastructure investment, like bridges falling down. Is the earth an organism that we are breaking? What will happen to the planet if we strip-mine the ocean floor? We don't know. Even if we knew the parameters, we don't have a computer that is big enough to simulate the system.

*Audience:*   Might not creating such radical fears about the environment do more harm than good?

*Joy:*   Anyone who isn't scared by now about what is happening to the ecology of the planet has got his or her eyes closed. We need to understand how this system works because we have already set massive changes in motion. We need to understand what the effects will be. Unfortunately the political system responds to noisy special interest groups, not to rational discussion, so the only way to get money focused on this problem is to make a lot of noise.

# 14

# Economic Consequences of New Technologies

Gordon Tullock

Throughout much of history people have worried about unemployment caused by improvements in efficiency. Although we do work less hard than we did before and take more leave, there is no evidence connecting improvements of efficiency with selective unemployment, that is, unemployment with individuals unemployed rather than everybody taking longer vacations. The reason is simple. When we have a gain in efficiency, we increase our total output of goods and services, hence we live better rather than holding consumption at the same level and unemploying people.

## New Technologies and Unemployment

There are two obvious economic consequences of any major improvement in technology. One is that we all get richer; the other is, a number of people say, that unemployment will increase. The emperor Claudius, in about 50 A.D., prohibited the use of a number of new building techniques on the grounds that he did not want to spread unemployment. The same ideas were expressed in the Middle Ages. Adam Smith's great book, *The Wealth of Nations,* starts with a discussion of a pin factory in the year 1776. This pin factory has 4 employees, and Smith points out that due to its advanced technology it was producing as many pins as could be produced by 400 people without this technology. So already in 1776, for every 4 people making pins there were 396 people unemployed.

We will return to the Industrial Revolution later; but going forward to the early days of the computer, a man named Ferry wrote a book,

*The Triple Revolution,* in which he said that the computer was going to unemploy substantially all clerical labor in the United States, and hence the government would need a major program to save them. In fact, even at that time, computers had already begun to *increase* the demand for clerical labor. It turns out that the reduction in the cost of producing clerical output as a result of using computers is so great, and the demand for clerical output is so elastic, that computers cause the demand for clerical labor to increase.

Most of the people who think that nanotechnology might lead to widespread unemployment would simply laugh at Emperor Claudius and Mr. Ferry, but they might have some difficulty explaining why their predecessors were wrong. The fear of unemployment is a long-standing fear. It is in fact an irrational fear if we consider the world population as a whole. (There are some special problems that I will return to in a moment.) The reason for this is very simple: we don't unemploy people, we just increase consumption. When things become easier to make, we use more of them. That turns out to be quite easy to do.

## Consumption

Most people can easily think of ways to increase their standard of living by 25 percent. If you ask them to consider a larger increase, they are at first a bit dubious, but if you point out the lifestyle of some of their neighbors, they quickly realize that they could increase their standard of living even further. If we increased world income ten times while keeping population constant, the average per capita income in the world would probably be slightly less than the average per capita income of those with advanced academic degrees. If we increased world income by a factor of 100—a large amount—it is likely that the average world per capita income would still be less than that of the average Ford dealer's. There is no reason that we cannot simply increase our consumption indefinitely.

Some people seem to think that we now have all we need, and we should not want anything more. But this is essentially a moral position and often implies that other people should not increase their consumption, but not the moralist. Those who take this position rarely reduce their own consumption. It is hard to say anything scientific about a view that morally people should reduce consumption—or at least not increase it—but it seems to me extraordinarily hard-hearted. With the exception of Africa, life expectancy in the so-called underdeveloped world has increased very sharply over the last few years. This is largely a result of increasing consumption.

## Advancing Technology and Flexible Skills

However, it is not true that every individual gains when total income rises. The reason that Andrew Carnegie came to the United States and became (for a while) the world's wealthiest man was that his father, a prosperous master weaver, was ruined by the last stages of the textile revolution. His father never recovered: he lived in the United States during the latter part of his life, trying to make a living by weaving handkerchiefs that he sold door-to-door. Fortunately, after a while his son was able to support him.

Another point illustrated here is that it was almost 70 years from the start of the industrial revolution—remember that it started with textile manufacturing—to the time when the weavers were destroyed. The destruction was very painful for the particular individuals who were wiped out, but it was not a sudden process.

Since about 1775 we have been living in a technological revolution. It may be going faster now, but it has been going on for over two centuries. One can anticipate that every major technological advance is going to damage people who have narrow specialized talents. Such people are not as common today as they used to be. Today, if you consider someone who is doing rather well in the economy, they are probably not skilled in the medieval sense of being able to do one thing very, very well. Rather, they are probably good at learning how to do *new* things well. We live in a rapidly changing environment, and there are not very many skills that one learns that remain useful indefinitely.

Individuals can be injured by technological advancement even in the face of a general increase in wealth. When cars from Japan and Korea became popular in the United States, the income of Ford dealers fell, except for those who were quickly able to get a Toyota franchise. Any improvement is likely to be opposed by some members of the community. We can say that, in the long run, the average person will be better off, but there are some who get hurt. If they can control the specific area of progress that is affecting them, and not stop the rest of progress, they may fare well.

## Politics and Regulation

I am particularly interested in the use of government regulations by individuals to increase their own wealth. It strikes me that this is actually the principal function of government regulations and that this is unfortunate. There are all sorts of things one would like to have regulated for the public good, but an examination of what we do in fact regulate shows that the bulk of regulations are an effort to increase the wealth of individuals. Through the U.S. farm program, the federal government spends on the order of $20 billion per year for the purpose of making food more expensive. It is very difficult indeed to see any argument for this except that a lot of farmers vote.

I am not arguing that it would be a good idea to get rid of all regulations. Presumably there are areas where a new technology like nanotechnology should be regulated, but I do not think we will know for a while what they are. My point is that if we begin to propose such regulations, the most likely ones are those that prohibit the use of some particular aspect of this technology because the income of people using the old technology would be lowered—although total income would increase—by the use of that aspect of nanotechnology.

This is not entirely one-sided. I am sure that those trying to introduce nanotechnology will also have their own expensive agents in Washington pushing the other way. (I should mention that the people who do this sort of thing in Washington are superior individuals; I just wish that they were doing something more valuable for society.) We normally expect a battle in Washington between those people who are attempting to get permission to produce a product that they expect will make a lot of money and other people who will be injured financially by the success of the first group.

The outcome of this battle is not very predictable in advance. As an example, consider the period in which the U.S. government ran a cartel in the airlines, from the time that the Civil Aeronautics Board was formed until the industry was deregulated. It was indeed true that the airlines developed rapidly, and it was even true that airfares were lower in the United States than they were in Europe, but it was not true that the industry developed as rapidly and as economically as it would have had the U.S. airlines not been regulated.

## Conclusions

We do not have to worry about the economic impact of increasing technology, such as nanotechnology—unless one feels that people should not be as wealthy as they are, and certainly not any wealthier. I live in Tucson, Arizona, a hundred miles from the Mexican border. The border between Arizona and Sonora is one of the places in the world where the living standard changes most abruptly. I cannot feel that the squatters living without running water on the Mexican side of the border really should be kept in that state. I would like to raise their standard of living. Thus I would like to have almost any kind of technological progress. I do not see any purely economic problems with nanotechnology, although we do not know enough yet to be sure. In any event, it certainly should not lead to widespread unemployment.

On the other hand, regulation and government control will probably be necessary, and if history is a guide, it will probably be handled very badly.

## Discussion

*Drexler:* Would you care to comment upon environmental quality as a form of wealth?

*Tullock:* It certainly is a form of wealth, and it is increasing. Agriculture is the principal activity that takes up ground and damages the environment. As a result of technological developments, we are growing more crops on the same number of acres. Since, to quote Adam Smith again, wealthy people don't consume more calories than the poor, this means that the total acreage under crops is going down, so that trees can come back, and are doing so quite rapidly in the United States as a whole. In the eastern part of the United States, areas that were under agriculture 50 to 100 years ago are now second-growth timber. Of course, second-growth timber isn't as desirable as climax forest, but wait another 50 years and we'll have really nice trees.

Air pollution is a clear case where government activity is required. It is equally clear that the government does a very bad job. For example, Tucson has a heavily subsidized bus line. The buses run around empty and as a result clearly increase the level of air pollution in Tucson rather than lowering it. Nevertheless, there is no doubt that with respect to both air and water quality, the unregulated market produces the wrong result. There are those who believe that the property rights system could be extended to improve air and water quality, but frankly I don't see how it can.

*Audience:*  If you assume that in the far future nanotechnology will be able to assemble virtually any manufactured good, replacing both costly materials and manufacturing labor, would you assume that the value of all other kinds of labor would go up astronomically?

*Tullock:*  The extreme case that you mention I would regard as unlikely in the near future, but I would think that we would certainly take more leisure. I don't think that the value of labor would go up; the value of products would go down. If it became so cheap to build a spacecraft and send it to Mars that anyone could afford it, I would anticipate that people would go.

*Audience:*  That smacks very much of the old prediction that nuclear energy is going to be so cheap we won't have to meter it. That was proven dramatically wrong. I would maintain that it is likely that if nanotechnology is successful it will *not* make things substantially cheaper than at present.

*Tullock:*  We would have to talk at length about what the inputs to the product are that you are talking about and how society is organized. My own guess is that things will become cheaper. It's been happening for the last 200 years; I know of no reason that the process of things becoming cheaper should stop.

# 15

## The Risks of Nanotechnology

Ralph Merkle

Molecular manufacturing systems and molecular nanotechnology will expand our abilities to make precisely the structures that we want and will do so at low cost. This will create a wide range of new opportunities and risks. It is hoped that the decisions we make and the actions we take in preparation for this technology will be wise and reasonable. I also hope that these decisions will be compatible both with human survival and with the flourishing of the human species within the natural order of things. This chapter discusses the risks and, in particular, the accidental risks inherent in the use of self-replicating systems as a basis for nanotechnology.

### Self-Replicating Systems

Nanotechnology has been described as the manufacturing technology of the twenty-first century for two reasons. First, it gives thorough control of the structure of matter. This control is obviously very desirable for all sorts of manufacturing. Second, nanotechnology provides this capability at low cost. One of the main reasons for low cost is the use of self-replicating systems. Self-replicating systems are an integral and important part of nanotechnology.

General-purpose self-replicating systems are unlikely to make mass-produced products in the future. Special-purpose manufacturing systems, finely tuned to the particular product that they are manufacturing, are more likely to be used to manufacture most, if not all, mass-produced goods. However, such special-purpose manufacturing systems must come from somewhere, and although they might be built by further special-purpose manufacturing systems, it seems that at some point you need a general-purpose manufacturing system to build the systems that build the systems that build the products. As an example, a metal lathe is a device that is normally not found in the home. However, it is used in industry, and even though it is not often used in the manufacturing process of a specific product, it is used to build the parts that go into the machines that are used to build things that eventually wind up in the home.

The initial cost to develop self-replicating systems will likely be quite high. It is difficult to estimate the time frame for developing self-replicating systems and to predict exactly what form self-replicating systems will take. Nonetheless, once we have them, costs will drop. In fact, costs should drop dramatically. Self-reproducing systems in nature are relatively cheap. Agriculture uses self-replicating systems to produce a large number of goods at fairly low cost. While currently still theoretical, the idea of using artificially designed and built self-reproducing systems is economically very attractive. At some point they should prove feasible.

Proliferation of self-replicating systems raise some concerns. The consequences are perhaps not immediate—which is good, for this will give us time to respond to the various concerns. Perhaps, in rather a change from the normal course of human affairs, we can begin to think about this technology and prepare for it before its actual arrival.

### Planning for Self-Replicating Systems

We should start thinking about this technology now for several reasons. First, there are imaginable scenarios that would be rather unpleasant. The downside risk in self-replicating systems, if done incorrectly, is arguably significant. Second, we need clear policies, regulations, and understandings as well as a general consensus on how to deal with such systems before implementing them. Once implemented, things may move quickly.

This point raises an obvious problem: a general consensus takes time to build. The development of laws, regulations, and policies requires long lead times, sometimes *very* long lead times. If you want laws, you have to initiate a legislative process. Before you can initiate a legislative process, you need a political process. Before the political process, you need education. Before education is possible, we need an understanding of the technology—and for that, we need thorough discussion of the technology. So far, we've just begun a serious discussion of the kinds of technologies we might have and the possible consequences they might bring.

Furthermore, international action is essential. Local regulations are fine for local problems. However, if you are planning to guide the development of technology across the entire planet, local regulations may be worse than useless. Local regulations can't cope with this kind of problem and would push technological development into less responsible locations.

Finally, there is a fair amount of misinformation in the culture about exactly how self-reproducing systems will develop. This point was driven home to me by a recent Star Trek episode. Star Trek tells us that "nanites" (as they are called) are very small and they are self-reproducing. They have an ability to survive in a wide variety of environments. They have the ability to evolve very rapidly (a remarkable ability in any system). They have a very complex social structure. Finally, they have good negotiating abilities. In fact, they get their own planet by the end of the episode.

Needless to say, with this kind of misinformation floating around, it behooves us to articulate the difficulties that might actually occur, what risks we might actually incur, and how we might avoid these risks.

As illustrated by the nanite episode, we (as a culture) have a number of misconceptions about self-replicating systems. For example, people normally think about *biological* self-replicating systems whenever self-replicating systems are mentioned. While this analogy is often informative, in actuality an artificial self-replicating system built sometime in the twenty-first century will probably bear little resemblance to any biological system. It's likely to bear as much resemblance to a biological system as a 747 bears to a duck. Nonetheless, we need to think about the problems involved in such artificial systems, and we should start thinking about them early on.

## Designing for Safety

The question is this. Can we safely design and use such systems? Barring deliberate abuse, can we design, build, and use self-replicating systems with some assurances that they will not cause problems? Can we in fact avoid accidental risks entirely? (This chapter does not address the serious but more political issues of deliberate abuse or the military applications of nanotechnology.)

We can state the problem as follows. Self-replicating systems pose two major risks. One is that a self-replicating system will continue to replicate unchecked. The other is that during replication there will be changes or alterations in the self-replicating system that will allow mutations that lead to some sort of evolutionary process. This could result in a system that does things that you don't expect—possibly quite unpleasant things that you don't expect.

Let us contrast biological systems and artificial systems, first considering energy usage. Biological systems use a variety of energy sources. A horse, for example, can munch on a variety of vegetables in nature. Bacteria can use many different sugars. By contrast, the artificial systems that we have built to date use fairly specialized energy sources. A car uses gasoline. The image of a car foraging for grass or munching on bushes is somewhat unlikely. There is reason to believe that artificially designed systems, unless they are deliberately designed so that they *can* forage in the wild, will have a hard go of it in a natural environment. In particular, if your artificial system is confined to utilizing a single source of energy, and if that single source of energy is presented to it in a relatively refined fashion (which appears to be an economically attractive alternative), cutting off that form of energy stops the system from functioning. Cars can't run without gasoline.

Second, there is a requirement for raw materials. Again, biological systems are quite capable of using a wide range of raw materials. Bacteria can utilize any of several different sugars combined with a few inorganic ions. Artificial systems today have many complex components. An automobile has lots of parts, and the economic way to build a car is to build some of the parts in the auto plant and some of the parts somewhere else. If you translate this principle to the design of an artificial self-replicating system, you can see that it is likely to be cheaper to have a system that gets at least some of its components handed to it on a silver platter. Unlike biological systems, which evolved in nature to survive in a harsh world, an artificial self-replicating system will be in a hothouse environment where it's fed particular resources designed for its specific needs. In particular, you can give your artificial self-replicating system compounds that are simply not found in nature but are relatively cheap and easy to produce. Not only does this make good economic sense, it's also a pretty good design requirement. I recommend against designing artificial self-replicating systems that are so robust that you could toss one into a fish pond and watch it swim away, thrive, and reproduce. This situation can and should be avoided in practice.

Another issue is the evolutionary capability of self-replicating systems. In nature, self-replicating systems evolve. In fact all the biological systems around us—from bacteria to people—evolve and are the result of millions of years of evolution. A major mechanism supporting this evolution is sex. Even bacteria have sex of a sort (they exchange DNA quite promiscuously). All higher organisms engage in sexual activities. This is a critical component in the ability to evolve. I would therefore suggest that we *not* include sex in artificially designed self-replicating systems. Some people object to this restriction, claiming that sex would be a great research tool, and that artificial evolution could do amazing things. It could indeed! If there are legitimate research needs for exploring these options, they should be done under tight constraints and should be approached with a great deal of caution.

Most people in the scientific and technical community are acquainted with the fact that artificial systems fail. Complex software systems, one particular variety of artificial systems, fail quite reliably. Even single errors can bring them down. It is possible and perhaps desirable to increase the probability that a single failure in a self-replicating system would cause it to fail utterly. For example, self-replicating systems must store instructions for building a copy of themselves somewhere. If these blueprints are encrypted properly, a single bit error would effectively erase the information. Such techniques can create systems that are even safer than simply constraining the input of energy and raw materials.

## Legislative Constraints

Conventional design techniques used for artificial systems do not lead to the kind of adaptability, flexibility, survivability, and evolutionary capabilities that you see in natural systems. But just because it is possible to design working systems that cannot survive in the wild and do not have the capabilities to evolve does not guarantee that systems with such capabilities will not be built. Consequently, you have to ask whether anyone would deliberately try to build a system that could survive in the wild and evolve, and consider how to persuade them not to. The pursuit of minor advantages in reckless disregard of the consequences could produce an unsafe system. Therefore, we need to develop a broad consensus that there are things that you do and do not do with self-replicating systems.

To summarize, we should *not* design self-replicating systems with

• The ability to evolve (including any kind of sexual inheritance mechanism)
• The ability to survive in a natural setting

Conversely, good design principals lead us to design self-replicating systems that require

• A single, highly refined fuel source
• Components not found in nature that the device cannot synthesize for itself

A combination of these approaches can provide us with self-replicating systems that cannot survive in nature and cannot "evolve" into something that could. To achieve this goal with a high degree of reliability will require some degree of regulation, consensus, and understanding on the part of everyone dealing with these systems. There are certain things you just *do not do*. Regulation during the current theoretical stage of the development of nanotechnology is not needed—nor would I propose such restrictions. However, before we can build any artificial self-replicating systems, well-designed international regulation will be essential.

**Discussion**

*Audience:* Can you give an analog of this type of control over a technology? You seem to be saying that the reason that we won't have these biological-like evolving systems is that we won't build them—even though we could. I cannot think of any historically equivalent withholding of a technology.

*Merkle:* I think that regulations already prohibit certain things in biotechnology. For example, you don't put botulinus toxin in E. coli.

*Audience:* We still aren't sure that these regulations will be followed over the long term. What I am asking is whether what you are suggesting is unprecedented.

*Merkle:* The artificial systems that we are going to be building need not bear any relationship to biological systems—and probably will not bear such a relationship if engineers are allowed to design economically attractive systems—unless we purposely try to copy the principles found in biological systems. Furthermore, there need to be global regulations, as well as a general consensus. People who want to copy biological systems, who want to steal components from biological systems that would make their devices risky, should be delicately informed that that is an unwise proposition.

*Audience:* Are you suggesting that there will be a top-down consensus?

*Merkle:* The ideal would be to develop consensus that led to appropriate international regulations. I don't think we can get international consensus on this without a great deal of effort—but local regulations simply won't work. For example, if a single nation were to get very concerned and put very tight controls into place, the result would likely be to drive researchers into another country. That does not effectively contain the problem, it simply shifts the geographic locale. As a consequence, it is important to develop an international consensus. Fortunately, I think we have time.

*Audience:* I sense a contradiction between your notion that we are not ever going to build nanotechnological devices to operate out in the real world and other ideas I've heard about using such devices to clean up toxic waste dumps or to use them inside human bodies.

*Merkle:*   Right! How do you take into account that we want to use nanotechnological systems in the real world? I think the answer is to divide up the functions. I don't think anything put into my body needs the ability to self-replicate and evolve. You can confine the self-replicating system to a tank in a factory with various requirements for exotic components. Those factory devices could then build systems that *would* be placed into my body—systems that would not replicate.

People often think of including all the necessary abilities embodied in a single system. Conceptually, this is the simplest approach: one widget that does everything. This gives you a lot of power, but it's risky. Instead, it is possible to divide the capabilities among many specialized widgets. One system can survive in the human body and perform cellular repair. Another system can self-replicate in a factory. A third system under lock and key in the laboratory could be capable of evolving and producing interesting variations that might be worthwhile for scientific research, and so on. Keep the abilities separate so that you never have one entity capable of doing everything.

*Schwartz:*   I am an engineer by education. I am often struck by the hubris of scientists and engineers in terms of the ability to control what we create. Earlier this year I had the opportunity to review a study carried out by the American Petroleum Institute on its future strategy. This was in late February and in it there was a sentence (before it was changed) that reads as follows: "Despite the fact that over the last decade we have experienced Bhopal, Chernobyl, Challenger, and Three Mile Island, we in the petroleum industry do not need to be concerned about major accidents that might cause problems for oil companies." This was three weeks before the Valdez ran aground. Why? Because they thoroughly believed that their measures were adequate for coping with any problem. Part of the problem is our ability to adequately anticipate real problems and to be able to deal with them when they arise. We imagine that we can control much more than we actually can.

# 16

# Fears and Hopes of an Environmentalist for Nanotechnology

Lester W. Milbrath

Evaluation of a technology unavoidably becomes an exercise in critical moral judgment. To start this chapter, I state my moral posture and invite readers not only to become more aware of theirs but also to recognize that development and deployment of a technology is an action requiring moral reflection and moral judgment.

The development of nanotechnology comes at a time when the basic thrust of our civilization must be deeply questioned. Serious critical analysis discloses that civilizations must be transformed. We must ask, "Will nanotechnologies aid in that transformation or will they impede it?" Nanotechnology will surely transform our lives in drastic ways; will that transformation lead to enhancement of values we cherish? Nanotechnologists must give serious and sustained attention to developing effective ways to control this powerful new capability—failing to do so would be a moral crime. The chapter concludes with a list of both fears and hopes regarding this new power that lies just beyond our reach.

## Basic Values

To evaluate a new technology we must attempt to see its consequences, which can only be evaluated within the context of the beliefs and values a person brings to the mental exercise. In this context, it is important to be self-conscious and open about one's beliefs and values. Doing so alerts the writer, and readers, to possible bias and can help to untangle misunderstandings and disagreements. It can also accelerate the process of social learning.

I begin by cryptically stating some of my core values (as best I can see them) that are most relevant to the concern of this chapter, so you can see "where I am coming from." The necessary brevity of this chapter requires cryptic statements without full exploration and justification. Those desiring to explore fuller reasoning behind these points should consult my book, *Envisioning a Sustainable Society: Learning Our Way Out.*[1]

Core values of this environmentalist:

1. The central value in my structure is *life in a viable ecosystem*. Note that I do not specify human life. Without a suitable habitat for other creatures, there is no suitable habitat for humans. Our species evolved in a flourishing, diverse, natural habitat, and we humans will always yearn for intimate contact with natural beings. Our natural self fits best into that niche from which we evolved. Our desire for life is rooted in our very nature; without that desire we would no longer exist. I cannot tell you why I want life; it is not instrumental to other values. It is basic.

2. I also want a *high quality of life*. Most succinctly, this means that I want the opportunity to realize all that I am capable of being. There are many other specific things that bring high quality to life, as I see it, but they need not be listed here.

3. My life cannot have high quality unless it is embedded in two flourishing systems: my *ecosystem* and my *social system*. I can imagine a flourishing ecosystem without any humans, but I cannot imagine a flourishing human society that is not embedded in a flourishing ecosystem.

These statements display a hierarchy of societal values. The top societal value—the top priority—is a flourishing, viable ecosystem. The second priority that a society must attend to is its own good functioning; the needs of the whole are more important than the desires of any individual. No life can have high quality unless it is embedded in a flourishing social system. A society can allow individuals to seek quality of life in any way they choose only so far as their actions lead to or do not contravene the good functioning of these two systems.

## Making Value Choices

We cannot escape moral choices; they always play a role in our decisions whether we are aware of them or not. They are more likely to be wise choices if we consciously keep our value priorities in mind. Our roles as scientists or technologists do not release us from our obligation to confront the morality of our labors. (A decision made *without* considering the consequences for our two basic systems is still a moral choice, even if it's not thought of in that light.)

All technologies affect the good functioning of these two systems. Therefore, all choices to develop or use a technology are moral choices that should be made with great care. Furthermore, technological impacts on systems are likely to be permanent. The more powerful or attractive the technology is, the greater and longer lasting its societal impact will be. (Witness the impact in just this past century of automobiles, television, and nuclear power.) Nanotechnology is potentially much more powerful and more attractive than any previously known technology. Choosing whether, when, and how to develop it could be the most important moral decision ever made by our species.

Technologies are akin to legislation—they shape how our society works, which, in turn, affects how our ecosystem works—but they are more permanent than legislation.[2] We can repeal a law, but we cannot repeal a technology. Ironically, we carefully debate (most of the time) the potential impact of a proposed new law, policy, or program, yet we never debate, and seldom try to anticipate, the impact of a new technology on our two vital systems.

The occasion for the presentation that became this chapter—the First Foresight Conference on Nanotechnology—is unique in that it is the first instance (to my knowledge) of an attempt to meet our moral responsibility to evaluate the impact of a potentially powerful and attractive new technology on our two vital systems. Figuring out how to control the pace and direction of nanotechnological development is one of the most urgent tasks facing our civilization. It is much more urgent than figuring out how to speed development of the technology.

Nanotechnology researchers need to build a pattern and tradition of discourse that self-consciously brings values into their analysis of plans for development and deployment. Any expenditure of effort inevitably assumes both societal values and personal values. What values are you assuming? What values do you want your work to serve? Does your work serve values that you do not want? As you think about these questions, can you anticipate the second-, third-, and even fourth-order consequences of your actions and inactions? Develop a habit of mind that continually asks, "And then what?"

If you could be completely successful in achieving everything you want to achieve with this new technology, what kind of world, what kind of life would we have then? What would be the impact on all of our other values? Would we have better justice? More freedom? More compassion? More honesty? Greater security? More fulfilling work? Better health? Greater peace? More equality? More control of our lives? Higher quality of life? It is not sufficient merely to say that we will have greater knowledge, or more goods, or longer lives, especially if achieving these diminishes other values. We must clarify the impact of a proposed achievement on our whole value package. Values are inextricably intertwined with each other and with the way society and nature work. We can never do merely one thing.

## A World without Humans

Before describing my specific fears and hopes for nanotechnological achievement, please join me in a thought experiment. Imagine that all humans vanish from planet Earth this instant, leaving behind all the structures, roadways, machines, ships, planes, and so on that make up our modern industrial civilization. Now allow several centuries to pass. What will have happened to all of our splendid human achievements? Entropy will have scattered most machines and structures. Plants and animals will have moved in and colonized the territory we pushed them out of. Nature will have settled down into a more balanced pattern. Many endangered species will have recovered a more viable population. Pollution, toxicants, and other poisons will be greatly diminished to assimilable levels. Biospheric and geospheric systems will have been able to recover their balance. All in all, nature will be flourishing splendidly again without humans.

I derive two truths from this thought experiment. First, there is a special wisdom in nature's well-established cycles and patterns. Nature as we see it today is what has been retained in good systemic working order after billions of failed experiments over billions of years. Our knowledge cannot fathom that complexity and intricacy. Ironically, we have somehow gained the power to disastrously disrupt these finely tuned systems, but we have not yet gained the social learning and wisdom to use our power in harmony with nature.

Second, we miss the point when we say that we have an environmental crisis. We have a crisis of the human species. Yet, humans lived in harmony with nature on this planet for more than two million years. It seems that the problem is not necessarily our physical makeup or with human culture. Rather, the problem is our specific civilization, our particular cultural variant. Our industrial civilization has taken a trajectory that cannot be sustained. Either we transform our civilization so that it can be sustained or nature will transform it for us. Nature's lessons are learned by pain and death—I fervently hope that this is not our destiny.

Increasing environmental degradation is on the verge of forcing the human species into massive social relearning. Such learning will be key to bringing about a societal transformation that could enable humans to live in a long-term, sustainable relationship with nature. As part of this social relearning we urgently need to learn such lessons as the following: (1) this would be a better world if there were fewer, not more, humans; (2) it is folly to speedily consume nonrenewable resources; (3) economic growth cannot be an enduring value; (4) a simple life can be rich and fulfilling even though we consume little; (5) compassion for other people, future generations, and other species is a vital life force; (6) partnership is better than domination; (7) cooperation is more successful than competition, (8) holistic, systemic, integrative thinking is more valid than linear, mechanistic thinking.

## Fears of an Environmentalist for Nanotechnology

My greatest fear is that this urgently needed relearning will be aborted if people perceive technological salvation to be just around the corner. Humans will not listen to messages for change so long as their systems seem to be working reasonably well. We are especially prone to look to new technologies to save our present way of life, rather than undergo a societal transformation that would have more enduring benefits. Development of nanotechnology, even the promise that it may soon be available, would forestall and probably foreclose this needed social learning. People would continue to believe in the current false gods of society: growth, consumption, wealth, competitiveness, power, and domination.

As an environmentalist, I hope that you nanotechnologists are not too "successful" too soon in developing this new technology. Give us more time to learn. Lend us your intellect and efforts for the most urgent task of relearning our way to a societal transformation.

Are you giving your intellect and time where it is most needed? By pursuing a new technology—which is tremendously exciting, and the rewards for the winner will certainly be great—there is a danger that in the excitement of the chase you will lose sight of other, more urgent problems. I speak specifically of the threat to life that is posed by drastic human-induced changes in biospheric systems. We must look *beyond* the expected changes in physical systems to anticipate the ramifications in social, economic, and political systems. Every effort must be made to forestall and mitigate these disruptive changes.

If you are a nanotechnological enthusiast, you are likely to say that developing these new technologies is the way to save us. I warn you that their development in time to avert drastic changes in biospheric systems is extremely unlikely. Developing artificial intelligence is critical to full-blown development of this new technological regime. Humans have been trying to do that for thirty years without much success. There is little prospect that you will be swiftly successful either.

Developing nanotechnology is every bit as difficult as turning back the forces that are leading to the destruction of global biospheric systems. But the latter is far more urgent. It is dangerously self-deceiving—and deceptive to the public—to seek resources for developing this new technology instead of working for the societal transformation that is necessary to save our species.

You have an urgent obligation to facilitate social relearning in another regard. K. Eric Drexler recognizes that nanotechnologies are so powerful that we must quickly learn to develop sociopolitical controls for their development and deployment.[3] Developing these controls is even more urgent than developing the technologies themselves. If someone develops and deploys the technologies before we can develop the necessary controls, we could well face the choice of becoming slaves to their domination or suffering massive physical and societal destruction. It is not good enough to say that "development of controls is someone else's job." *It is your job.* Developing a powerful new technology without having first developed satisfactory sociopolitical controls is morally wrong. It is a crime against all life.

## Hopes of an Environmentalist for Nanotechnology

As an environmentalist, I do not believe that a nanotechnological revolution will come in time to avert the catastrophe we foresee. Nor do I have confidence that its benefits would outweigh its costs should it arrive. On the other hand, I know that we cannot stop nanotechnological development. Rather than waste energy opposing it, we will use our scarce time and resources to seek societal transformation that is much more likely to lead to a better life. Obviously, I think it would be better for those of you who are working to develop these technologies to join us, but I realize you are likely to press ahead with development despite my warnings. What guidance can I give for your journey?

Nanotechnology moves beyond the heating, beating, shaping, and fastening of our modern bulk technologies—the technologies that teach us power and domination and which nature cannot readily accommodate. Nanotechnology emulates nature and could be deployed much more harmoniously with it. Could nanoresearchers accept the time-honored wisdom of nature and strive to develop structures that emulate and fit in with natural structures, thereby making them supportive of nature? Can you avoid structures and processes that depart radically from nature or that seek to remake nature?

If this stance were to become the posture of the research community, could the search for nanotechnological understanding foster a deeper appreciation of the elegant complexity and intricacy of nature? Could the knowledge we will acquire be used to help people better perceive, accept, and appreciate a loving bond with nature's awesome elegance and beauty? On the other hand, if you foster a nanotechnological revolution that leads people to be more distant from nature and to dominate other creatures ever more powerfully, this will be fought by nearly all environmentalists.

With that caveat, I will list some of the ways that nanotechnologies could help mitigate some of the presently perceived urgent problems resulting from human degradation of our environment:

1. Could nanomachines be used to extract carbon from the atmosphere, thereby diminishing the greenhouse effect? Could the extracted carbon be combined with natural cellular structures to make humus, which could in turn be used to rebuild depleted soils?

2. Could nanomachines seek out and breakdown ambient chlorofluorocarbons before they reach the stratosphere and destroy the stratospheric ozone layer? Could they rebuild the ozone layer or build a similar ultraviolet-screening device?

3. Could nanotechnologies transform transportation so as to make it light, efficient, safe, and nonpolluting? Such a development would reduce carbon-dioxide buildup, conserve scarce nonrenewable resources, and reduce toxic injuries to living creatures.

4. Could nano*dis*assemblers seek out and invade toxic dumps to breakdown complex chemical structures into naturally occurring elements? Could similar devices seek out and breakdown toxicants in air, water, and soils?

5. Could nanodisassemblers and reassemblers analyze the genetic structure of preserved specimens of extinct species and bring them back to life? Could they be wise enough to regenerate the habitat that once nourished those species?

I urge you to use your talents to help humans learn how to regenerate natural systems and to recover sustainability rather than build human-built systems as substitutes for nature. If you set out to conquer or dominate nature, not only do I believe you will fail but I, and many other environmentalists, will impede your efforts. I hope you direct your efforts and pace your development so that we can work together.

### References

1. L. Milbrath, *Envisioning a Sustainable Society: Learning Our Way Out* (Albany, New York: SUNY Press, 1989).

2. L. Winner, *The Whale and the Reactor: A Search for Limits in an Age of High Technology* (Chicago: Chicago University Press, 1986).

3. K. E. Drexler, *Engines of Creation* (Garden City, NY: Anchor Press/Doubleday, 1986).

# 17

## The Weapon of Openness

Arthur Kantrowitz

"The best weapon of a dictatorship is secrecy, but the best weapon of a democracy should be the weapon of openness."
—Niels Bohr[1]

What is the "weapon of openness," and why is it the best weapon of a democracy? Openness here means public access to the information needed for making public decisions. Increased public access (i.e., less secrecy) also gives information to adversaries, thereby increasing their strength. The "weapon of openness" is the net contribution that increased openness (i.e., less secrecy) makes to the survival of a society. Bohr believed that the gain in strength from openness in a democracy exceeded the gains of its adversaries, and thus openness was a weapon.

This is made plausible by a Darwinian argument. Open societies evolved as fittest to survive and to reproduce themselves in an international jungle. Thus the strength of the weapon of openness has been tested and proven in battle and in imitation. Technology developed most vigorously in precisely those times—the industrial revolution—and precisely those places—western Europe and America—where the greatest openness existed. Gorbachev's glasnost is recognition that this correlation is alive and well today.

Let us note immediately that secrecy and surprise are clearly essential weapons of war and that even countries like the United States—which justifiably prided itself on its openness—have made great and frequently successful efforts to use secrecy as a wartime weapon. Bohr's phrase was coined following World War II when his primary concern was with living with nuclear weapons. This chapter is concerned with the impact of secrecy versus a policy of openness on the development of military technology in a long-duration peacetime rivalry.

Let us also note at the outset that publication is *the* route to all rewards in academic science and technology. When publication is denied, the culture changes toward the standard hierarchical culture in which rewards are dependent on finding favor with superiors. Reward through publication has been remarkably successful in stimulating independent thinking. However, in assessing openness versus secrecy, it must be borne in mind that research workers (including this author) start with strong biases favoring openness.

In contrast, secrecy insiders come from a culture in which access to deeper secrets conveys higher status. Those who "get ahead" in the culture of secrecy understand its uses for personal advancement. Knowledge is power, and for many insiders access to classified information is the chief source of their power. It is not surprising that secrecy insiders see the publication of technological information as endangering national security. On the other hand, to what degree can we accept insiders' assurances that operations not subject to public scrutiny or to free marketplace control will strengthen our democracy?

My own experience relates only to secrecy in technology. Therefore I will not discuss such secrets as submarine positions (which seem perfectly justifiable to me in the sense that they clearly add to our strength) or activities that are kept secret to avoid the difficulties of explaining policy choices to the public (which seem disastrously divisive to me).

First, we offer some clues to understanding the historical military strength of openness in long-duration competition with secrecy. Second, we suggest a procedure for utilizating more openness to increase our strength.

## The Strength of Openness

An important source of support for secrecy in technology is the ancient confusion between magic and science. In many communications addressed to laypeople the terms are used almost interchangeably. Magic depends on secrecy to create its illusions; science depends on openness for its progress. A major part of the "educated" public and the media have not adequately understood this profound difference between magic and science. This important failure in our educational system is one source of the lack of general appreciation of the power of openness as a source of military strength. A more general understanding of the power of openness would bolster our faith that open societies would continue to be fittest to survive.

Openness is necessary for "the processes of trial and the elimination of error," Sir Karl Popper's beautiful description of the mechanism of progress in science.[2] Let's try to understand what happens to each of these processes in a secret project, and perhaps we can shed some light on how the peacetime military was able to justly acquire its reputation for resistance to novelty.

*Trial* in Popper's language means receptivity to the unexpected conjecture. There is the tradition of the young outsider challenging the conventional wisdom. However in real life it is always difficult for completely new ideas to be heard. Such a victory is almost impossible in a hierarchical structure. The usual way a new idea can be heard is for it to be sold first outside the hierarchy. When the project is secret this is much more difficult, whether the inventor is inside or outside the project.

Impediments to the elimination of errors will determine the pace of progress in science as they do in many other matters. It is important here to distinguish between two types of error that I will call ordinary and cherished errors. Ordinary errors can be corrected without embarrassment to powerful people. The elimination of errors that are cherished by powerful people for prestige, political, or financial reasons is an adversary process. In open science this adversary process is conducted in open meetings or in scientific journals. In a secret project it almost inevitably becomes a political battle and the outcome depends on political strength, although the rhetoric will usually employ much scientific jargon.

Advances in technology incorporate a planning process in addition to the trial and elimination of error that is basic to all life. When the planned advance is small, the planning can be dominant, in the sense that little new knowledge is required and no significant errors must be anticipated. When the planned advance is large, it will usually involve research and invention, and the processes of trial and the elimination of error will determine the rate of progress. In these cases the advantages of openness will be especially important. The familiar disappointments in meeting schedules and budgets are frequently related to the fact that, in selling new programs, the importance of these unpredictable processes is not sufficiently emphasized. More openness would reduce these disappointments.

Trial and the elimination of error is essential to significant progress in military technology, and thus both aspects of the process by which significant progress is made in military technology are sharply decelerated when secrecy is widespread in peacetime. Openness accelerates progress. For peacetime military technology, openness is a weapon. It is one clue to the survival of open societies in an international jungle.

### Secrecy as an Instrument of Corruption

The other side of the coin is the weakness secrecy fosters as an instrument of corruption. This is well illustrated in President Reagan's 1982 Executive Order #12356 on national security (alarmingly tightening secrecy), which states {Sec. 1.6(a)}: "In no case shall information be classified in order to conceal violations of law, inefficiency, or administrative error; to prevent embarrassment to a person, organization or agency; to restrain competition; or to prevent or delay the release of information that does not require protection in the interest of national security."[3]

This section orders criminals not to conceal their crimes and the inefficient not to conceal their inefficiency. But beyond that it provides an abbreviated guide to the crucial roles of secrecy in the processes whereby power corrupts and absolute power corrupts absolutely. Corruption by secrecy is an important clue to the strength of openness.

One of the most important impacts this corruption from secrecy has is on the making of major technical decisions. Any federally sponsored project, and especially a project so hotly contested as the Strategic Defense Initiative, must always keep all its constituencies in mind when making such decisions. Thus the leadership must ask itself whether its continual search for allies will be served by making a purely technical decision one way or the other. (A purely technical decision might determine whether money flows to Ohio or to Texas. Worse yet, revealing technical weaknesses could impact the project budget.)

When this search for allies occurs in an unclassified project, technical criticisms, which will come from the scientific community outside the project, must be considered. Consideration of these criticisms can improve the decision making process dramatically by bringing a measure of the power of the scientific method to the making of major technical decisions.

In a classified project, the vested interests that grow around a decision can frequently prevent the questioning of authority necessary for the elimination of error. Peacetime classified projects have a very bad record of rejecting imaginative suggestions that frequently are very threatening to the existing political power structure.

When technical information is classified, public technical criticism will inevitably degrade to a media contest between competing authorities and, in the competition for attention, it will never be clear whether politics or science is speaking. We then lose both the power of science and the credibility of democratic process.

Corruption is a progressive disease. It diffuses from person to person across society by direct observations of its efficacy and its safety. The efficacy of the abuse of secrecy for interagency rivalry and for personal advancement is well illustrated by the array of abuses listed in Sec. 1.6(a). The safety of the abuse of secrecy for the abuser is dependent upon the enforcement of the section. As abuses spread and become the norm, enforceability declines and corruption diffuses more rapidly.

However, diffusive processes take time to spread through an organization, and this makes it possible for secrecy to make a significant contribution to national strength during a crisis. When a new organization is created to respond to an emergency, as for example the scientific organizations created at the start of World War II, the behavior norms of the group recruited may not tolerate the abuse of secrecy for personal advancement or interagency rivalry. In such cases, and for a short time, secrecy may be an effective tactic. The general belief that there is strength in secrecy rests partially on its short-term successes. If we had entered World War II with a well-developed secrecy system and the corruption that would have developed with time, I am convinced that the results would have been quite different.

## Secrecy Exacerbates Divisiveness: the SDI Example

Reagan's executive order, previously referred to, provides another clue to the power of openness. The preamble states: "It [this order] recognizes that it is essential that the public be informed concerning the activities of its Government, but that the interests of the United States and its citizens require that certain information concerning the national defense and foreign relations be protected against unauthorized disclosure."

The tension in this statement is not resolved in the order. It may be informative to attempt a resolution by considering a concrete example, namely the Strategic Defense Initiative. SDI symbolizes one of the conflicts, clearly exacerbated by secrecy, that currently divide us.

There are unilateral steps toward openness that we could take, and would leave us more unified and stronger, even if no reciprocal steps were taken by the nonaligned nations. I propose that we start unclassified research programs designed to provide scientific information needed for making public policy. If these programs are uncoupled from classified programs, their emphases would not compromise classified information. Their purpose would be to provide a knowledge base for public policy discussions. These programs would not reveal the decisions taken secretly, but a public knowledge base would reduce the debilitating divisiveness fostered by secrecy.

The Strategic Defense Initiative provides a classic example of debilitating divisiveness. Countermeasures to SDI are deeply classified. The deadly game of countermeasures and counter-countermeasures will probably determine whether SDI is successful or a large-scale Maginot Line. At the present time, classification of the countermeasure area trivializes the public debate to a media battle between opposed authorities offering conflicting interpretations of secret information.

An example of this game is decoying versus discrimination. If the offense can proliferate a multitude of decoys that cannot be discriminated from warheads by the defense, SDI will not succeed. Knowing a decoy design would of course make it easier for an adversary to discriminate it from a warhead. It is therefore very important that such designs be carefully guarded. On the other hand, maintaining secrecy concerning the scientific and engineering research basic to the decoy-discrimination technology would, for the reasons discussed earlier, make it much more difficult to provide assurance to the public that all avenues had been explored. Indeed, a substantial part of the criticism of the feasibility of SDI turns on the possibility that an adversary would invent a counter-measure for which we would be unprepared.

### The Cryptography Case: Uncoupled Open Programs

We can learn something about the efficiency of secret versus open programs in peacetime from the objections raised by Admiral Bobby R. Inman, former director of the National Security Agency (NSA), to open programs in cryptography. NSA, which is a very large and very secret agency, claimed that open programs conducted by a handful of mathematicians around the world, who had no access to NSA secrets, would reveal to other countries that their codes were insecure and that such research might lead to codes that even NSA could not break. These objections exhibit NSA's assessment that the best secret efforts that *other* countries could mount would miss techniques that would be revealed by even a small uncoupled open program. If this is true for other countries is it not possible that it also applies to us?

Inman (1985) asserted that: "There is an overlap between technical information and national security which inevitably produces tension. This tension results from the scientists' desire for unconstrained research and publication on the one hand, and the Federal Government's need to protect certain information from potential foreign adversaries who might use that information against this nation."[4]

I assert that uncoupled open programs (UOP) in cryptography make America stronger. They provide early warning of the capabilities an adversary might have in breaking our codes. There are many instances in which secret bureaucracies have disastrously overestimated the invulnerability of their codes. In this case I see no tension between the national interest and openness. The cryptographers have provided a fine case study in strengthening the weapon of openness.

Consider then the value of starting unclassified, relatively inexpensive, academic research programs uncoupled from the classified programs. These UOPs could provide the more solid information on countermeasures needed for an informed political decision on SDI, just as the open cryptography research has taught us something about the security of our codes. If indeed SDI's critics are right about the opportunities for the invention of countermeasures, then the UOP would provide an opportunity to make a conclusive case. On the other hand, if the open programs exhibited that SDI could deal with all the countermeasures suggested and retain its effectiveness, its case would be strengthened.

These open programs would indeed be shared with the world. They would strengthen the United States even if there were no response from unfriendly nations by reducing corruption by secrecy, by improving our decision making, and by reducing our divisiveness. Undertaking such programs would exhibit our commitment to strengthening the weapon of openness. Making that commitment would enable democratic control of military technology. More openness, reducing suspicions in areas where Americans are divided, will do more to increase our military strength by unifying the country and its allies than it could possibly do to increase the military strength of its enemies.

### The Weapon of Openness and the Future

Bohr's phrase, which is the keynote of this chapter, was coined in an effort to adapt to the demands for social change required to live with advancing military technology. Unfortunately Bohr's effort, to persuade FDR and Churchill of the desirability of more openness in living with nuclear weapons, was a complete failure. There can be no doubt that the future will bring even more rapid rates of progress in science-based technology. I mention three possibilities, noting that these are only foreseeable developments, and that there will be surprises that, if the past is any guide, will be still more important.

1. Artificial intelligence is advancing, driven by its enormous economic potential and its challenge in understanding brain function.
2. Molecular biology and genetic engineering are creating powers beyond our ability to forecast.
3. Some years ago, Feynman wrote a paper entitled, "There's Plenty of Room at the Bottom" [see appendix B], pointing out that miniaturization could aspire to the huge advances possible with the controlled assembly of individual atoms. When the possibility of the construction of assemblers that could reproduce themselves was added by Eric Drexler in his book, a very large expansion of the opportunities in atomic-scale assem-

bly were opened up.[5] This pursuit, today known as nanotechnology, will also be driven by the enormous advantages it affords for health and for human welfare.

But each of these developing technologies has possible military uses comparable in impact to that of nuclear weapons. With the aid of the openness provided by satellites and arms control treaties, we have been able to live with nuclear weapons. We will need much more openness to live with the science-based technologies that lie ahead.

## Notes

1. N. Bohr. Personal communication.
2. K. Popper, *Logic of Scientific Discovery* (New York: Harper & Row, 1968).
3. Presidential Executive Order #12356, 1982, {Sec. 1.6(a)}.
4. R. Inman, "Technology in East-West Relations," *Technological Frontiers and Foreign Relations*, National Academy of Sciences, National Academy of Engineering, and The Council on Foreign Relations (Washington, DC: National Academy Press, 1985).
5. K. E. Drexler, *Engines of Creation* (Garden City, NY: Anchor Press/Doubleday, 1986).

# 18

## What Public Policy Pitfalls Can Be Avoided in the Development and Regulation of Nanotechnology?

*Panel Discussion:* K. Eric Drexler, Arthur Kantrowitz, Ralph Merkle, Lester Milbrath, Peter Schwartz (moderator), and Gordon Tullock

*Audience:* Two factors that at least superficially favor closed (i.e., nonpublic) development of nanotechnology are (1) the return on investment for industry, and (2) national concerns regarding the transfer of technology. For example, Japan has the beginnings of a development program for these kinds of technologies; the United States may or may not develop such a program. Could you comment on how you see these issues affecting the global development and diffusion of nanotechnology?

*Kantrowitz:* One of the great developments that made the industrial revolution possible was the evolution of the patent, which opened up inventions in exchange for a brief monopoly on their exploitation. We need to find ingenious ways to expand our protection of intellectual property so that it will be possible to develop these technologies in the open.

*Audience:* Given that secrecy is a dangerous thing for a generally open society, do you have any idea what the impact would be of eliminating it completely? Everyone seems to take for granted the idea that government secrecy is important for national defense. I am not sure that this has been proved.

*Merkle:* Some areas of secrecy are quite useful. For example, there are personal records that one would not like to have distributed too widely, like the results of urinalysis in a drug-testing program. As for secrecy in the interest of national defense, there was the comment that secrets are quite appropriate in narrowly defined areas.

*Kantrowitz:* You have to distinguish between war and peace. When you are at war, you have to have secrecy.

*Schwartz:*  I was recently visiting with Yevgeny Velikhov, the chief scientist for Gorbachev. He thought that one of the most important things that they could do was to get satellite dishes into every school and every apartment building in the [former] Soviet Union. Then the revolution would be unstoppable because secrecy could no longer be preserved. That is a very different environment from trying to prevent the Nazis from invading Stalingrad.

*Audience:*  With regard to the previous comments on patent policy, I would like to ask about secrecy in corporations of the United States versus the situation in Japan. In Japan, there is much more communication between scientists in different corporations. The Hokai system allows you to park an idea, along with the date, in the patent system without actually having a patent. In the United States, because of corporate secrecy, we lack a mechanism for large-scale corporate cooperation.

*Tullock:*  That is because our patent system doesn't work right. The idea of the patent system is to make secrecy unnecessary by giving the inventor intellectual property. In fact, you are required to publish in order to get a patent. Unfortunately, if I were in a large company, I would not trust the patent system and would use secrecy. We could improve the patent system.

*Kantrowitz:*  Let me give a different answer to that question. There is a difference in the organization of corporations in the United States and in Japan that doesn't help this country. They still have a great many powerful companies run by engineers. Ours are run by lawyers and accountants. Lawyers and accountants do not understand the processes of trial and the elimination of error, which are the mechanisms of technological progress as well as science.

*Audience:*  I want to dispute the notion that what we need to do in order to help the poor is to improve technology. What we need to do is to put money into their hands. Then we would find that the technology that we already have is more than adequate to generate the wealth that is needed. Today the output of our overall economy is limited primarily by the distribution of wealth, and the ownership of capital is the fundamental issue. When you introduce advanced technologies that are even more productive, you will find that, yes, costs do drop, but wages scale the same way. It is unpredictable what the outcome of new technology will be, but it is certainly plausible that we will end up with a situation where everyone is out of a job and one or two people who happen to have stock in the right companies will own everything. That is the reductio ad absurdum that nanotechnology presents to us of a process that has been underway for at least the last hundred years.

*Tullock:* Over the last fifty years, the poorest people in "The United World," living in places like Africa and India, have about doubled their life span. They have benefited from the technology that we have developed. They have not benefited as much as they might have had we been more generous and given away large parts of our wealth, that's true. Most people are willing to give away about five percent of their wealth, but not much more.

*Drexler:* I will exercise a panelist's prerogative to ask another panelist a question. Ralph, your remarks made it sound as if with nanotechnology we will have sufficient control that we won't have to worry about accidents. Are these indeed your views?

*Merkle:* I was discussing the extraordinary accidents that are made possible by the concept of a self-replicating system. I think that there will continue to be the ordinary sort of accidents in which someone simply makes a mistake and produces, for example, a bad design for a medical machine that runs around inside the body and kills someone instead of healing them. I am not aware of any way to prevent ordinary accidents.

The kind of accident that I was discussing is the kind that would have an extraordinary scale because of the properties of a self-replicating system made possible by nanotechnology. The objective was simply to point out that it should be feasible, with due caution, preparation, and forethought, to avoid this class of accident entirely and to design systems that simply would not have the capability to do things that we regard as dangerous in the narrow sense of unchecked replication, and in particular unchecked replication coupled with mutation and evolution.

*Drexler:* So we could still have problems of the Bhopal, India, and Exxon, Valdez sort, as Peter suggested?

*Merkle:* Yes.

*Milbrath:* The assumption in Ralph's solution is that we will have a planetary consensus, and that this consensus will be enforced all over the globe. We have yet to develop a planetary politics, which must precede a planetary consensus. My estimate of the best way to do this is to develop planetary consensus regarding saving planetary systems.

One of the best ways to work for planetary politics is to try to help people all over the world develop an understanding that these are questions that require consensus. Nanotechnology has got to be controlled, and if it is not controlled, it will be abused in some way and injure all of us. This is a danger that lies in the failure of politics rather than any danger inherent in [Dr. Merkle's] solution. This technology could be achieved in other parts of the world by people who have no consideration for us, so that we all run the danger of being enslaved by it. The only way to avoid the slavery is to become planetary politicians.

This is especially important for the people at this conference, because science reaches across national borders more easily than any other community. We can get a response to the necessity of this kind of control within the scientific community, across national boundaries. It is especially incumbent on you as scientists to reach out to others in other countries to try to develop this consensus. This is how we will build a planetary politics, and it is your responsibility because no one else can do it as well.

*Merkle:* I would like to add a few points to that. I think that the development of international cooperation in the development of a new technology is particularly attractive. Nanotechnology offers a number of advantages for this kind of development. It is far enough away that I don't think it will be subject to any regulation of communication, so that you can exchange views on this subject with friends in foreign countries and discuss it without being accused of giving away vital secrets. Nanotechnology is not within the short-term planning horizons of most companies, so that it is possible to discuss nanotechnology internationally today and to develop international ties now. With any luck, by the time people realize that there is money to be made, we will have a strong international community with a strong set of beliefs about continued international cooperation that will survive the tensions that develop with commercial interest.

*Drexler:* I would like to emphasize that it is not just possible today in the metaphorical sense, but it is possible today in the literal sense, given the composition of this conference.

*Kantrowitz:* I'd like to say, however, in view of the striking results of this conference, that it won't be long before an appreciation of nanotechnology arrives—and an awareness of its power. This will put an end to the very pleasant openness that exists in nanotechnology today.

*Merkle:* In that case, we had best communicate quickly.

*Birge:* The point that I would like to make is that what we are calling nanotechnology is generally recognized in Japan to be of major importance. They are putting enormous resources into it. They are moving ahead of us at a significant rate, and we are going to be playing catch-up with Japan and probably Germany for the next five years. We are not going to be calling the shots here at all. We are going to be trying to make up for time lost due to the apparent inability of American industry to think more than five years ahead.

I am very critical of American industry because I have interacted with a number of companies and they have said "Come back when you have something that we can do in two years." I have said that I would be back in three years, but it would still take five more years to develop; meanwhile the Japanese and the Germans are moving ahead. Unless we have a policy dictated either by the government or by a more expansive research program in industry, we are not going to be calling the shots in this country at all.

*Merkle:* Let me add that that is another excellent reason for international cooperation.

*Kantrowitz:* I would say also that that would not prevent us from imposing secrecy in order to pretend to be in the lead, just as we did with superconductivity.

*Schwartz:* Just to support your remarks, Dr. Birge, among the 18 leading American and European corporations that are corporate members of the Global Business Network, only one of them is represented here. Having spoken with the heads of R&D of all of them, every single one—even the one that is represented here—described nanotechnology as science fiction, as something not really worth paying attention to.

*Tullock:* The sensible thing to do, if the Japanese or the Germans are putting a lot of time and money into research, is to parasitize on their research. You can save money and get the same results. I admit that that means that you [Dr. Birge] won't get a research contract, but from the standpoint of the average American consumer, it's better. I think that parasitism is a better word than cooperation.

*Birge:*    That's stupid. In a field that moves as rapidly as this, by the time that we've made something following their lead, they're way ahead of us again.

*Tullock:*    We stay two years behind them and have no research budget.

*Birge:*    And we will make nothing. We're doing that in television sets now.

*Audience:*    I'd like to talk about sex because, according to studies, we all think about it a lot, but also because it is an important issue. In a previous incarnation I was a biologist, and I learned respect for a number of things including: (1) evolution occurs very fast, especially among bacteria, and (2) microbes have been evolving rapidly for 4.5 billion years, with generation times often as short as 45 minutes. Any plan to generate nanotechnology that doesn't take full advantage of the experience of the biosphere is probably neglecting a very good resource for learning.

I have a second comment, also on the subject of sex. There have been two recurring themes that have struck me in this conference. One is self-replicating devices. Another is information transfer. Information transfer is a major component of sex. I don't think that you can have information transfer, which is the goal of many of these nanomechanisms, and self-replication, and not have sex. Therefore, you will have evolution, and that must be dealt with.

*Merkle:*    Certainly I think that it is a good idea to steal as many ideas as possible from biological systems. I am merely in favor of making certain that the systems that you build don't start building upon those ideas themselves. As far as the idea that information exchange must necessarily lead to alterations in the blueprints of the devices themselves, I think that is clearly false. I certainly would not want the software inherent in the self-replicating devices that are free in the environment to have the ability to use information to modify their blueprints. That ability is an additional level of software that I would not want to see in systems free in the environment.

*Drexler:* Biological systems evolved to evolve. They are not really partitioned and modular the way that engineering systems can be. Just as one can run software on a computer without modifying the hardware of the computer, likewise one could have nanomechanisms that do exchange information without themselves being modified by that information, because that information is not blueprint information. It is in a different domain.

*Audience:* I don't think that you can separate the transfer of information that is required to achieve self-replication from the transfer of information that is the "intended goal." The experience in biology ten years ago is relevant here. People were going to whip up a batch of crippled microbes for use in doing genetic engineering, so that if the microbes were released either deliberately or inadvertently, a fail-safe mechanism would prevent those microbes from surviving in the environment. The genetic engineers found that this was a much harder exercise than was anticipated. The microbes that learn how to evolve are the ones that succeed.

*Drexler:* "Learning to evolve" requires evolution. If you are not able to take a single evolutionary step, you cannot evolve, or get better at evolving. Ralph pointed out, using his background in cryptography, that you can design replicators in which changing a single bit is the functional equivalent of erasing the genome. It is very hard to learn to evolve when the first change leads to erasure. That is really a very solid result. It is entirely unlike biology.

*Audience:* Wouldn't that require a universal agreement to have this encryption scheme built into every self-replicating mechanism?

*Drexler:* It wouldn't have to be a single universal scheme; there are many equally effective approaches. Indeed, even if you don't do something equivalent to genome encryption, there is every reason to believe that it does indeed require having "evolved to evolve" to be effective at evolving. If you make modifications to the parts of a machine, the parts don't match anymore; they haven't evolved to evolve. Organisms are much more flexible. You can have Siamese twins in which the organism still functions even though evolution did not select for surviving that aberration. Biological systems have a truly amazing flexibility and adaptability, and to believe that we are going to achieve that flexibility *accidentally* is fortunately not reasonable.

*Audience:*  You're doing two things here. You are saying that we don't have to worry because self-replicating systems couldn't evolve without having been carefully designed to do so. Then Ralph is saying that we have to do the equivalent of having a copy-protection scheme built into every device to make sure the systems don't evolve.

*Drexler:*  The fact that there are two mechanisms, either one of which makes you completely safe, should not make you feel less safe. The human imagination is such that you often come up with more than one solution even when only one is sufficient. Coming up with the second does not mean that the first was incorrect; it just means that you kept thinking.

*Merkle:*  I must confess that as a cryptographer I am prone to cooking up ideas along those lines whether or not they are particularly necessary. I would also say, addressing an earlier concern, that modifying existing biological systems, as in the case of making crippled bacteria, is taking a system that is already designed to be quite robust and attempting to make it less robust. This should be contrasted to the de novo design of a self-replicating system where you should very carefully not do all kinds of things which would make the system robust in certain senses.

*Kantrowitz:*  I think that this discussion of the mechanisms for preventing disastrous evolution is premature. We are also suppressing something that we all know, that the power of evolution will put an enormous premium on its use in creating powerful nanotechnology. Thus, it will not be easy to enforce the regulations that we have debated today.

*Audience:*  You have to consider the possibility of the creation of life. What is the panel's view of the ethical and theological consequences if we do in fact create life?

*Merkle:*  As long as the nanomachines are not conscious, I think that we are okay. If you start talking about the design and creation of conscious entities, then you enter a new area that we have not been addressing at this conference. So far we have been discussing building devices that are very small and have a lot of raw computational power. The field of artificial intelligence [AI] is, I think, dedicated to the idea that you can use whatever hardware is available to build a device which is perhaps conscious. That is a separate debate and one which I am happy to leave to the artificial intelligentsia.

*Audience:* In AI, in fact, this pressure to look at techniques like evolution has already led to areas like genetic algorithms and various learning architectures, and those seem to be the right architectures, as far as we can tell, for building flexible robots. If you want a powerful, useful nanorobot, as far as we can tell, on the computing side, you will use learning mechanisms.

*Merkle:* I am definitely in favor of using evolutionary mechanisms as long as you do it inside a computer and I get to pull the plug.

*Audience:* But the computer is hooked up to the hand on the nanomechanism because you need such a link for a learning mechanism.

*Merkle:* You are basically saying that necessarily there is a need to mix the computational power and the ability to explore algorithms through algorithm space, on the one hand, with a device capable of doing something obnoxious to lovable me, on the other hand. I would like to say that I would like those capabilities to be separated. If you have a robot arm hooked up to an evolving, wonderfully intelligent software program, my response is that I don't want that software hooked up to that arm.

*Audience:* That is the direction in which we are moving to get flexible robots. We don't know how to do them otherwise.

*Milbrath:* This discussion leaves me very uneasy. I don't think that it is going to be easy to say that we will draw a line and not cross it, especially if we have to get worldwide consensus on it.

*Audience:* These discussions of consequences and approaches are all very important, but we have to face the truth that in the long run, someone will pay somebody enough money to do something that will be destructive. That is what happens. There are hydrogen bombs on the floor of the ocean that shouldn't be there. Mistakes happen. We have to think not only about how to prevent one of our miscalculations from creating a problem, we have to consider how to fix someone else's accidental or deliberate mistake. We need something like an environmental antibody that we can use if we have to.

*Merkle:* I would like to make one clarification. I was talking about accidental errors. The whole area of deliberate abuse of the technology is another set of issues.

*Drexler:*    Yes. I would like to separate out a few conceptual categories which I think are important in the mental maps of the people who have been thinking about this for a while. In the area of nanotechnology, there are some very real concerns associated with the development of a technology of great power in a world of diversity, a world without any centralized decision-making body. That makes it very difficult to decide *not* to develop the technology. It makes it difficult (but *somewhat* less difficult) to decide how to guide the technology if it is being developed. I believe that the latter course, although it is difficult, as has been emphasized here, is the course that we must follow.

In considering what aspects of the development of nanotechnology to worry about, given that the technology is being developed, one concern is how to draw lines such that staying on one side of the line makes extraordinary accidents very unlikely. Another concern is how to avoid crossing those lines once drawn. I believe that it is clear, as Ralph has explained, that there is a line that can be drawn such that if you stay within it you can do almost everything that you are motivated to do— in terms of constructing devices that would be useful in the world—with no chance of an extraordinary accident. On the other hand, there are some incentives to cross that line. I think that those incentives can be reduced by drawing the line carefully so that people can do as many things as possible and still have them be safe. Ralph's emphasis on keeping different capabilities separated so that you don't combine dangerous capabilities in the same devices is a step in the right direction.

We have many years to think about how to give people as much scope as possible for experimentation and beneficial applications while avoiding the possibility of extraordinary accidents. We need to spend those years thinking and building a consensus on policies that will minimize the chance that that line will be crossed. That chance will nevertheless remain. That chance concerns me very deeply. But I don't want to see us worrying too much about concerns that are on the safe side of the line, when our big problem is drawing a good line and keeping from crossing it.

*Kantrowitz:*    It seems to me that this discussion has been a little unbalanced in neglecting the whole world of military competition. Nanotechnology is going to play a role in that, unless we can succeed in abolishing military competition. That is a line that could be drawn.

*Drexler:* Yes.

*Audience:* Professor Milbrath mentioned several times a vast social transformation that he thought was necessary, but he never said what it was. I'd like to ask him what he thinks such a transformation might be.

*Milbrath:* I did very briefly discuss some characteristics of the transformation. I've just written a 400-page book about it, *Envisioning a Sustainable Society.* I think that we need to rethink many characteristics of modern society. If we could develop the kind of society that I want, I think we could eliminate both war and the military, and the kinds of dangers that we talk about here. But that is a total societal transformation, and one that I am reasonably sure that nature will force us into. Whether we can be wise enough to go through it in such a way that we come out with something intact, I don't know. I am trying to help that process. I do not believe that modern society, as presently constituted, can continue. The trajectory that it is on is unsustainable in numerous ways, which I discuss in my book. That is my challenge to all of you. Think about your society. It is more important than what you are thinking here. What you are thinking here is only a small but dangerous part of the whole problem.

*Kantrowitz:* I'd like to tell you about a phrase that for me characterizes this meeting very well. I read it originally in Dawkins's book, *The Blind Watchmaker.* He uses the phrase "linkage disequilibrium," and I think that we have achieved it. This meeting, having very wisely gathered people from all kinds of disciplines, has achieved a kind of intellectual ferment that is analogous to what happens when ecologies that have been kept separate suddenly are allowed to mingle. You get a burst of evolution. I look forward to witnessing this evolution.

*Schwartz:* Thank you, Dr. Kantrowitz. I'll just add my closing remarks in the same spirit. One of my favorite essayists is a man named Paul Valéry, the poet laureate of France during the mid 1930s and 1940s. He wrote an essay on unpredictability and the role of science. One of the most happy phrases in that famous essay seems particularly appropriate for the conclusion of this meeting: "The future isn't what it used to be."

# Appendix A: Machines of Inner Space

K. Eric Drexler

We live in bodies made of atoms in a world made of atoms, and how those atoms are arranged makes all the difference. To be healthy is to have tissues and cells made of correctly patterned sets of atoms. To have wealth is, in large measure, to control collections of atoms organized to form useful objects—whether foodstuffs, housing, or spacecraft. If we could arrange atoms as we pleased, we would gain effectively complete control of the structure of matter. Nanotechnology will give us this control, bringing with it possibilities for health, wealth, and capabilities beyond most past imaginings.

This sort of dominion over nature will not arrive overnight, but only after years of hard work in various enabling technologies. Nonetheless, examples from chemistry and nature already show many of the basic possibilities.

Chemists find that atoms can bond to form molecules. Every diamond is a single, huge molecule made of carbon atoms (figure A.1); every breath of air contains pairs of nitrogen and oxygen atoms, each pair a small molecule. Molecules, whether large or small, are objects. Each has such properties as size, shape, mass, strength, and stiffness. Nanotechnology will use nanometer-scale molecular objects as components of molecular machines.

Nature shows that molecules can serve as machines because living things work by means of molecular machinery. Enzymes are molecular machines that make, break, and rearrange the bonds holding other

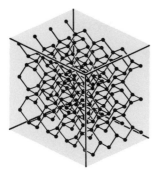

**Figure A.1**
Diamond: one cubic nanometer, 176 atoms.

molecules together. Muscles are driven by molecular machines that haul fibers past one another. DNA serves as a data storage system, transmitting digital instructions to molecular machines (ribosomes) that manufacture protein molecules. And these protein molecules, in turn, make up most of the molecular machinery just described. Nanotechnology will exploit a similar strategy, using programmable molecular machines—called *assemblers*—to build things, including more molecular machines. Assemblers will work like tiny industrial robots, directing chemical reactions by positioning molecular tools to build complex structures atom by atom (figure A.2).

Microtechnology enables construction on a scale of micrometers, or millionths of a meter. Nanotechnology will enable construction on a scale of nanometers, or billionths of a meter. The term *nanotechnology* is sometimes used (especially in Britain) to refer to any technology giving some control of matter on a nanometer scale. This definition would seem to include glass polishing, the manufacture of thin films, and ordinary chemistry. As used here, however, the term describes a technology giving nearly complete control of the structure of matter on a nanometer scale. Since atoms are themselves about a third of a nanometer in diameter, this sort of nanotechnology will require a general ability to control the arrangement of atoms.

Even this control, however, will not give us what the alchemists sought—a way to turn lead into gold. In essence, nanotechnology will be a vast elaboration of ordinary chemistry and biology. It will move

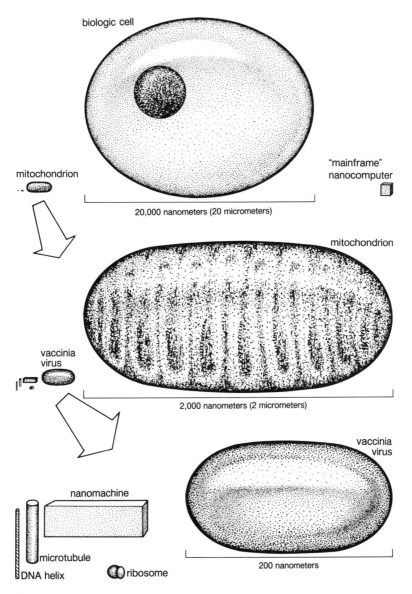

**Figure A.2**
Microscopic components of the biological world and artificial molecular devices compared at successive levels of magnification.

atoms around, as in chemical reactions, not fuse them or split them, as in nuclear reactions. To transform one element into another will be beyond its ability.

How might we bridge the gap between present abilities and nanotechnology? Two very different paths are being pursued today: one is a top-down strategy of miniaturizing current technologies; the other is a bottom-up strategy of building ever more complex molecular devices, atom by atom.

## The Top-Down Approach

The top-down strategy sees the problem as one of scale, of pushing back the frontier of miniaturization. This is the tradition of the watchmaker and the manufacturer of integrated circuits. This century has seen great progress in miniaturization, from the shrinking of clockwork into fingernail-sized boxes to the shrinking of computers onto fingernail-sized chips—and this progress continues.

One of the great visionaries of the top-down strategy was physicist Richard Feynman. In 1959, he gave a talk in which he proposed that large machines could be used to make smaller machines, and those to make machines still smaller, working step by step toward molecular dimensions. He envisioned microscopic lathes, and described problems with building microscopic automobiles for mites. But at the end of his talk, he turned to the molecular size scale and hinted at the need for a bottom-up approach. "The principles of physics, as far as I can see," he said, "do not speak against the possibility of maneuvering things atom by atom." He went on to remark, "But it is interesting that it would be, in principle, possible (I think) for a physicist to synthesize any chemical substance that the chemist writes down. Give the orders, and the physicist synthesizes it. How? Put the atoms down where the chemist says, and so you make the substance." Nevertheless, Feynman suggested that these substance-synthesizing machines "will be really useless," because chemists will be able to make whatever they want without them.

Modern microtechnology has followed a different path, also discussed by Feynman in his talk. Microtechnologists use beams of light and electrons to make patterns on surfaces, and they use techniques such as selective chemical etching and the deposition of thin films of metals, oxides, and semiconductors to develop those patterns into structures. The patterns can be microns or fractions of a micrometer across, and the structures can be wires and transistors, or even mechanical devices.

Microtechnology for electronics is an old, yet fast-developing story; it has given us modern computer technology. Microtechnology for machines is more recent and remains in the research phase. Researchers have deposited patterned films of silicon over films of oxide, then dissolved the oxide to release silicon parts. Using clever patterns of deposition, Kaigham Gabriel and William Trimmer of AT&T Bell Labs have shaped interlocking silicon gears, trapped against the surface by silicon flanges but left free to rotate. Streams of air can spin these gears like turbines at over 15,000 revolutions per minute. Richard Muller at the University of California at Berkeley has made the first electrostatic motors using a similar technology.

Silicon micromachining is pushing back the frontier of miniaturization, but how close has this top-down approach come to nanotechnology? In scale, it remains vastly different. Microgears and micromotors are now tens of microns in diameter, but nanogears and nanomotors will often be tens of nanometers in diameter, or less—a thousandth the linear dimension and a billionth the volume of current micromachines. This is analogous to the difference between a truck and an integrated circuit.

More fundamental, however, is the difference in quality of construction. Micromachining, whether with microlathes or with etched patterns, can only shape materials from the outside. It cannot build them from the inside, from the bottom up, and so it cannot give complete control of the structure of matter. These top-down technologies have many uses (consider microcomputers), and they may even give us nanoscale devices, but they cannot evolve into true molecular nanotechnology.

## The Bottom-Up Approach

In contrast with the top-down strategy, the bottom-up strategy sees no problem with making things small; chemists and biochemists already make small molecules with ease and in abundance. From this perspective, the problem is to make things large, while keeping detailed, molecular control of structure.

**Table A.1**    Macroscopic and molecular components

| Technology | Function | Molecular examples |
|---|---|---|
| Struts, beams, casings | Transmit force, hold positions | Cell walls, microtubules |
| Cables | Transmit tension | Collagen, silk |
| Fasteners, glue | Connect parts | Intermolecular forces |
| Solenoids, actuators | Move things | Muscle actin/myosin |
| Motors | Turn shafts | Flagellar motor |
| Drive shafts | Transmit torque | Bacterial flagella |
| Bearings | Support moving parts | Single bonds |
| Clamps | Hold work pieces | Enzymatic binding sites |
| Tools | Modify work pieces | Enzymes, reactive molecules |
| Production lines | Construct devices | Enzyme systems, ribosomes |
| Numerical control systems | Store and read programs | Genetic system |

Adapted from K. E. Drexler, *Proc. Nat. Acad. Sci.* 78 (1981): 5275–78.

The bottom-up strategy was originally inspired by chemistry and molecular biology. For over a century, chemists have understood molecules as small three-dimensional objects to be broken down and built up. In 1926 physicist Erwin Schrödinger supplied the foundations for a quantum mechanical theory of molecules and chemical bonding. In 1944 he wrote a book, *What is Life?*, which correctly viewed life as based on molecular objects and machines. In 1953 James Watson and Francis Crick determined the three-dimensional structure of DNA, and four years latter J. C. Kendrew and his colleagues in England determined the three-dimensional structure of the first protein. Since then, molecular biologists have detailed the structures and functions of molecular devices at an ever increasing rate.

It was the existence of a wide range of natural molecular machines (see table A.1) that led me to propose artificial molecular machines, ultimately including such things as molecular assemblers, assembler-based replicators, mechanical nanocomputers, and cell repair machines. Molecular machines in nature showed that a bottom-up approach to nanotechnology would work. Indeed, the most clearly workable bottom-up approach begins by mimicking nature, by designing new protein-based devices.

Proteins are polymers made by joining many smaller molecules—amino acid monomers—to form chains. The monomers of protein chains form specific sequences, like the letters in a specific sentence. The sequence of monomers in a protein likewise has a special significance: it determines how the chain will fold, coiling back on itself to form a compact three-dimensional object. A folded protein can be as stiff as a piece of wood or engineering plastic. Furthermore, depending on its shape, surface properties, and so on, a folded protein can do things; for example, it can serve as an enzyme, as a structural element, or even as part of a molecular motor. Just as a protein chain can fold up to form an object, so can a collection of proteins stick together to form a larger, more complex object, thus enabling the construction of complex molecular machines.

Nonmolecular machines may also be of use in nanotechnology. In 1982, Gerd Binnig and Heinrich Rohrer at IBM's research laboratory in Zürich, Switzerland, announced their development of a device called the scanning tunneling microscope, or STM (figure A.3). Like its younger relative, the atomic force microscope (AFM), the scanning tunneling microscope can move a sharp tip over a surface with atomic precision. This capability immediately suggested a modified bottom-up strategy for developing nanotechnology, using the STM tip to position molecular tools for precise molecular construction.

In the long term, the approach used to reach nanotechnology will make no difference because the early, clumsy technologies will swiftly be left behind. In the short term, however, the rates of progress in different approaches will be decisive, determining not only which approach pays off but how fast nanotechnology itself arrives.

### Building with Molecules

Various bottom-up approaches have produced experimental results. Bare STM tips—without molecular tools—have made molecular-scale modifications on surfaces. At Bell Labs, Murray Hill, New Jersey, R. S. Becker, J. A. Golovchenko, and B. S. Swartzentruber have produced what appear in STM images as atom-sized bumps on germanium crystal surfaces,

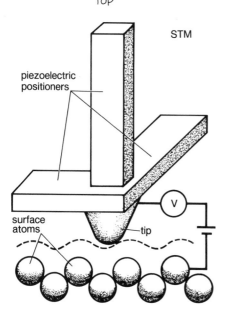

**Figure A.3**
Schematic for scanning tunneling microscope.

apparently produced by evaporating single atoms of germanium from the tip of an STM. At IBM's Almaden Research Center, San Jose, California, J. S. Foster, J. E. Frommer, and P. C. Arnett produced bumps on graphite crystals, chemically pinning fragments of single molecules to the crystals' surfaces using current from an STM tip. The presence and absence of such bumps might be used one day to store the ones and zeros of binary coded computer data, crowding many trillions of bits into a square millimeter.

Thus far, however, such experiments have failed to produce specific, controllable molecular changes; the detailed nature of the bumps has been unpredictable. It seems unlikely that bare STM tips, even those ending in a single atom, can provide the precise, molecular control needed for building nanomechanisms. Whether or not nanotechnologists use STM or AFM devices for positioning, they will likely need molecular tools to construct the first generation of nanomachines.

In the long run, nanotechnology will probably use robot-like molecule-sized assemblers to position molecular tools, but assemblers are not necessary to begin building with molecules. Instead, one can use self-assembly to form larger structures from molecular components suspended in solution, using principles familiar to chemists and molecular biologists. Even assembler-style positioning using STM or AFM tips, if the technique becomes useful, will probably begin by using a self-assembly process to cap those tips with molecular tools.

Self-assembly differs radically from ordinary manufacturing techniques. It involves small pieces joining together automatically to form larger objects. Making a molecular device this way will be much like growing a crystal, whereby a solid three-dimensional object is built up by adding layer after layer of molecules to its surface. Whereas crystals have simple, regular structures made of only a few kinds of molecules, molecular machines will be complex, irregular structures made of many different kinds. The component molecules themselves could be made by chemical synthesis, by mixing reactive compounds together in the right order under the right conditions.

Chemical synthesis and self-assembly have real advantages over atom-by-atom positioning. Using these methods to synthesize even a thousandth of a gram of a typical protein (a modest amount, by current standards), results in more than a million billion ($10^{15}$) molecular components, without the need for performing some billion billion ($10^{18}$) molecular assembly operations. In making a gram of devices, each self-assembled from a thousand molecular components, one would again make $10^{15}$ molecular objects and save another $10^{18}$ individual assembly operations. To pull cost numbers out of the air, at a millionth of a cent per assembly operation, this would mean a savings of $10 billion. Also, unlike alternatives, this kind of synthesis and assembly is already known to work in the laboratory and in nature.

How, though, can self-assembly work for molecules when it does not for cars or computers? The key principles are selective stickiness and thermal motion. Proteins and other large molecules can have complex shapes and surface properties. Two such molecules can fit like pieces in a jigsaw puzzle, having not only complementary shapes, but complementary patterns of attractive forces; for example, electrical charge. They will stick to each other, but not to molecules that lack this complementarity—this is selective stickiness. Warm molecules bounce about (which is the reason the molecules in air do not fall to the floor and stay there). In a liquid, these thermal motions affect everything, including large molecules and molecular devices suspended in solution. Thermal motions have little effect on large objects like automotive and computer components, but they can bring molecular components together in all possible positions and orientations. Selective stickiness can then cause self-assembly.

Self-assembly can be swift and effective. In a solution, thermal motions shift a typical protein by its own diameter every millionth of a second and twist it to a substantially different angle in a ten-millionth of a second; fine motions (of an atomic diameter or so) happen many billions of times per second. These constant motions allow molecules to "explore" their environments quickly, finding any complementary molecules and sticking to them.

Nevertheless, to make self-assembling machines, one must first make their parts. Nature often uses proteins as parts, each containing hundreds or thousands of atoms. Modern advances in chemistry and genetic engineering make new proteins reasonably straightforward to produce. The challenge, however, has been to design protein chains that fold up correctly to form solid molecular objects with the desired properties.

In recent years, this challenge has been taken up by many groups, including those led by David and Jane Richardson of Duke University, Durham, North Carolina, Bruce Erickson of the University of North Carolina, and David Eisenberg of the University of California at Los Angeles (UCLA). William DeGrado of the DuPont Company, Wilmington, Delaware, and Robert Hodges of the University of Alberta have shown some of the first successes. Indeed, it has proved possible to design proteins more stable than those found in nature—proteins that can better serve as building blocks for engineering design. The design techniques needed to make a single protein fold correctly are much like those needed to make separate proteins self-assemble to form a larger structure. Thus, these successes in protein design mark a major milestone on the self-assembly path to nanotechnology.

Proteins show promise enough, but there are alternatives. In 1987, Donald J. Cram of UCLA and Jean-Marie Lehn of Louis Pasteur University, Strasbourg, France, and the Collège de France, Paris, received the Nobel Prize for chemistry (shared with United States researcher Charles J. Pedersen) for synthesizing relatively simple molecules that perform functions like those of natural proteins, selectively binding other molecules. Such molecules could be designed to self-assemble, providing an alternative to proteins for building molecular machines. Myron L. Bender of Northwestern University, Evanston, Illinois, and Ronald Breslow of Columbia University, New York City, have made nonprotein molecules that function as enzymes. These general structures are easier to design than proteins because their three-dimensional shapes do not depend on a complex folding process. The catch is that they are often hard to synthesize, and hard to scale up.

An alternative strategy, not yet explored, would be to use folding polymers much like proteins but to build them from a set of molecular subunits chosen to simplify synthesis and folding. This would sacrifice a key advantage of proteins (the convenience of biological production) but would combine some of the ease of design offered by general structures with some of the ease of synthesis offered by polymers.

Any of several bottom-up strategies using self-assembly seem feasible. The question is not which can be made to work, but which can be made to work first, with the lowest costs or the highest payoff. Today's technology is already working with molecular devices; the challenge is to make them larger, more complex, and more capable.

### Goals along the Way

With or without the long-term goal of assemblers and nanotechnology, researchers are already following bottom-up strategies motivated by short-term payoffs. Scientists have designed folding proteins to answer scientific questions. Protein engineers have built enzymes more stable than their natural counterparts so they will last for months in detergent bottles on store shelves. They aim to build new enzymes to help make industrial chemicals and pharmaceuticals. Other molecules will be engineered to serve as chemical sensors or as detectors for medical diagnostics. Each step will hone skills in the design and fabrication of molecular objects.

Electronics has inspired yet more ambitious goals. Microelectronics has followed a top-down strategy of miniaturization, but bottom-up researchers are mounting a challenge under the rubric of "molecular electronics." From early work by Avi Aviram of IBM's Thomas J. Watson Research Center, Yorktown Heights, New York, and Forrest Carter of the U.S. Naval Research Laboratory, Washington D.C., to current work by such researchers as Richard Potember of Johns Hopkins University, Baltimore, Maryland, Robert Birge of Carnegie-Mellon University, Pittsburgh, Pennsylvania, Kevin Ulmer of seQ, Ltd., Cohasset, Massachusetts, and Mark Wrighton of the Massachusetts Institute of Technology, interest has grown. The first applications will likely involve piling up large numbers of molecules having special electronic properties to make quantities of new electronic materials. Over the longer term, many researchers hope for nothing less than molecular electronic circuitry in which individual molecular parts can serve the roles of wires and transistors.

A molecular electronic circuit or perhaps an entire computer could be made by self-assembly, possibly using proteins as scaffolding. Picture a snap-together puzzle, with each piece having a transistor or a piece of wire glued to it. The puzzle pieces link together to form a circuit board, and their interlocking shapes determine just how the electronic components join up to form circuits. Self-assembly of molecular electronics could work this way, with large molecules as the puzzle pieces and attached electroactive molecules as the components. Unlike the essentially two-dimensional integrated circuits of today, however, these puzzle pieces would go together to form a three-dimensional block of circuitry.

Interest in molecular electronics has been strong in Japan, where it is a focus of the international Frontier Research Program of RIKEN (the Institute of Physical and Chemical Research). Interest has also been high in the Soviet Union and Eastern Europe, where it is seen as a possible way of leapfrogging Western semiconductor technology.

Before entire computers can be made by self-assembly, similar techniques will probably enable the construction of moderately complex molecular machines. A machine only a few times more complex than a typical enzyme (and far simpler than a ribosome) could help synthesize special folding polymers, perhaps including copies of itself. Using tools to help build better tools is an ancient story in engineering; it is at the heart of most progress.

It will eventually make sense to construct assembler-like devices in order to help build molecular structures that are impossible to make through self-assembly. These protoassemblers (whether purely molecular or a self-assembled hybrid of molecular tools and an AFM- or STM-style positioning mechanism) would have only a limited ability to position a limited range of tools, but they could be used to build better tools. This process will culminate in general assemblers: molecular machines that can serve as powerful engines of creation, opening the era of true nanotechnology.

### Assemblers and Nanotechnology

Assembler-based nanotechnology will give nearly complete control of the structure of matter, enabling the construction of the smallest, strongest, fastest devices possible under natural law as it is understood today. The road to nanotechnology may pass through protein technology, using parts as stiff as wood, but nanotechnology itself will use materials like diamond, four times as stiff as steel and fifty times stronger. Parts made of such materials cannot be made by mixing ordinary chemicals or by coaxing large molecules to self assemble; they will be made through direct control of chemical reactions, using robot-like assemblers wielding reactive molecules as tools.

Crude, early assemblers will be made by protoassemblers. Later assemblers, however, will be made using advanced assemblers, which means that they themselves can be made of strong, rigid materials like diamond. To picture such a device, one should imagine not a protein molecule or a biological cell but a jointed, computer-directed industrial robot arm, full of gears, bearings, and drive shafts, yet only a ten-millionth of a meter long (figure A.4).

Assemblers can do their jobs using reactions like those familiar to chemists. What they will add is control: whereas a chemist mixes molecules in solution, letting thermal motions bring them together haphazardly (and often with unwanted side effects), an assembler arm will bring molecules together with precision, making a single reaction occur in a specified location. Enzymes show this sort of control, but they are specialized for particular jobs; assemblers will tackle general construction tasks.

The uncertainty principle of quantum mechanics places some limits on how accurately a molecular tool can be positioned, but at ordinary temperatures thermal motions pose a greater problem. The solution is

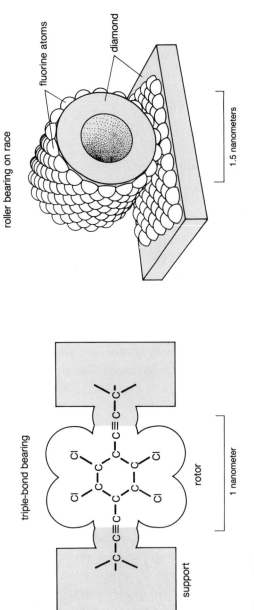

**Figure A.4**
A molecular rotor is supported by two bearings made of carbon-carbon triple bonds (left). A roller bearing on a race with fluorine atoms forming bearing surfaces for structural elements of diamond (right).

to make the arm stiff enough to hold its position with sufficient accuracy. This will require careful design, but calculations show that it can work.

Today, what humans can do is strongly limited by what they can build. It would be straightforward to design a billion-bit computer memory chip, but no one bothers because no one can build such a thing. By contrast, assemblers will enable construction of almost anything having a chemically-reasonable arrangement of atoms. Then what mankind can do will be limited not so much by fabrication abilities as by cleverness and by the limits inherent in natural law.

## Small Products

In nanomachines, a single atom can serve as a gear tooth, and a single bond between two atoms can serve as a bearing. Whole arrays of atoms sliding over one another can serve as stronger bearings in larger devices. Drive shafts, cams, roller bearings, levers—a full range of familiar mechanical devices—can be built on a nanometer scale (figure A.5). Standard electromechanical devices, such as electromagnetic motors, will not work on this scale, but while electromagnetic motors lose power in small sizes, other motors gain power. Calculations indicate that direct-current electrostatic motors just 50 nanometers in diameter can provide mechanical power at a rate of hundreds of trillions of watts per cubic meter; that is, many nanowatts per motor. With motors and mechanical parts, complex systems will become possible, including assemblers and computers to control them.

At this stage, the study of nanotechnology rests on what may be termed exploratory engineering, involving studies of what can be built with tools that do not yet exist. Exploratory engineering differs from science in that it aims merely to apply existing scientific knowledge, not to uncover new facts about nature. It differs from standard engineering in that it attempts only to make the case that certain future systems are (or are not) physically possible, not to actually design and build those systems today. The discipline of exploratory engineering uses both analogy and calculation. The above case for assemblers uses analogies from biology and chemistry; the case for molecular computers, however, must rely more on calculations.

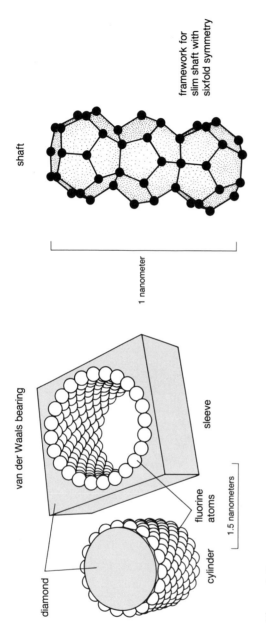

**Figure A.5**
Fluorine atoms cover the mating surfaces of a cylinder and sleeve made of diamond to form a van der Waals bearing (left). Drive shaft (right).

From the perspective of standard engineering, molecular electronic computers will likely prove best, but from the perspective of exploratory engineering, they pose problems. Their operation involves the complexities of molecular quantum mechanics, raising difficulties of design, modeling, and calculation. This gives reason to consider what can be done with molecular mechanical computers. Although likely to be far slower than molecular electronic computers, they have the virtue of being easy to analyze with classical physics. Their projected performance can set lower bounds on what future computers can do.

Where today's electronic computers move signals by putting a voltage on a conductor, a mechanical nanocomputer would do so by displacing a molecular rod. In an electronic computer, signals on one conductor can control signals on another by means of transistors. In a mechanical nanocomputer, signals would control one another through mechanical locks. Thermal motions will tend to disrupt these operations, but again the problem can be solved by building parts with adequate stiffness (figure A.6).

Mechanical systems are intrinsically slower than electronic systems of comparable size, but smaller systems operate proportionally faster. Cutting size in half doubles speed (the moving parts have less far to travel), and a mechanical nanocomputer with a central processing unit a fraction of a micron across would be moderately faster than current electronic computers. Moreover, a half trillion of them (with space for cooling channels and interconnections) could fit in a cubic centimeter, putting more computing power in a desktop machine than exists in the entire world today.

Raw computer power does not translate directly into intelligence, for without suitable software the most powerful computer could not even add a column of figures. But raw power must be some help. It is worth noting that current artificial intelligence research has been trying to coax human-like behavior out of computers with roughly a "microbrain" of computing power; that is with a *millionth* that of the human brain. With nanocomputers a desktop machine could have a raw computational power level in the "megabrain" range. This capability might make a difference.

Single nanocomputers running conventional software (like that used in automated factories) can be used to operate assemblers. Given fuel,

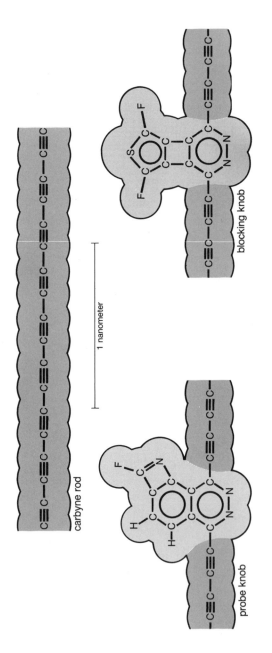

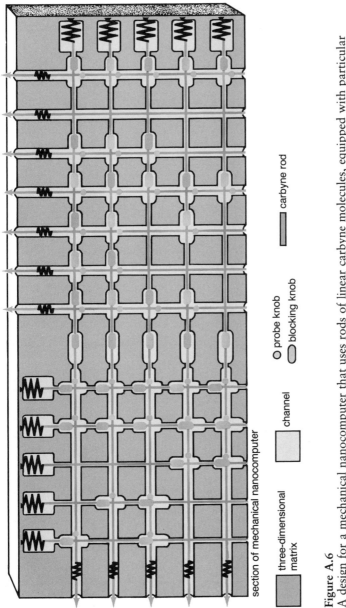

section of mechanical nanocomputer

- three-dimensional matrix
- channel
- ○ probe knob
- ◯ blocking knob
- ▬ carbyne rod

**Figure A.6**
A design for a mechanical nanocomputer that uses rods of linear carbyne molecules, equipped with particular molecular configurations serving as blocking and probe knobs, all sliding within channels of a three-dimensional matrix.

raw materials, and the right instructions, assemblers will be able to make virtually anything—including more of themselves. This will make possible *replicators*, nanomachines able to copy themselves in vats of suitable chemicals. Calculations suggest replication times of about a thousand seconds, letting a single, microscopic replicator give rise to many tons of product replicators in a day or so. After reprogramming, the product replicators would team up to build other things, again by the ton.

## Big Consequences

It should come as no surprise that self-replicating molecular machines can build big, and at low cost. After all, this is the way redwood trees come into existence. And like plants, nanomechanisms will be able to use solar power, but at least an order of magnitude more efficiently. With efficient solar-electric energy converters as inexpensive as grass, and with strong, tough diamond-fiber composite hardware as inexpensive as wood, much will become possible. Inexpensive fuel and efficient spacecraft, for example, should eventually make space flight less expensive than air travel is today.

Parallels with other products of natural molecular machinery suggest further applications. For example, plants gave Earth its oxygen atmosphere and created the carbon found in coal and oil. Today, people fear that rising atmospheric carbon dioxide levels from the burning of fossil fuels may overheat the Earth through the greenhouse effect. If the solution does not come first in some other way (perhaps by the planting of more trees), solar-powered nanomechanisms could reverse the carbon dioxide buildup, taking a few years of operation to turn all the excess carbon dioxide back into carbon and oxygen.

Nanomachines, with their broad ability to rearrange atoms, will be able to recycle almost anything. Using nothing but sunlight and common materials, and with no by-products other than waste heat, they will produce a wide range of products. With production costs similar to those of plants, they will enable the clean, rapid production of an abundance of material goods. The benefits could be especially dramatic for the third world.

Potential medical applications also show that small systems can have big effects. Cells and tissues in the human body are built and maintained by molecular machinery, but sometimes that machinery proves inadequate: viruses multiply, cancer cells spread, or systems age and deteriorate. As one might expect, new molecular machines and computers of subcellular size could support the body's own mechanisms. Devices containing nanocomputers interfaced to molecular sensors and effectors could serve as an augmented immune system, searching out and destroying viruses and cancer cells. Similar devices, programmed as repair machines, could enter living cells to edit out viral DNA sequences and repair molecular damage. Such machines would bring surgical control to the molecular level, opening broad new horizons in medicine.

While contemplating the potential benefits of nanotechnology, though, one would be well advised to spend time contemplating its potential harm. The chief danger is not likely to be that of runaway accidents; replicators, for example, need no more have the ability to function in a natural environment than an automobile has the ability to refuel from tree sap. Deliberate abuse is another matter. One need only consider the prospect of programmable "germs" for biological warfare to see the seriousness of the problem.

Nanotechnology will let humankind control the structure of matter, but who will control nanotechnology? The chief danger is not a great accident but a great abuse of power. In a competitive world, nanotechnology will surely be developed. If democratic institutions are to guide its use, it must be developed by groups within the reach of those institutions. To keep nanotechnology from being wrapped in military secrecy, it seems wise to emphasize its value in medicine, in the economy, and in restoring the environment. Nanotechnology must be developed openly to serve the general welfare. Society will have years to shape policies for its beneficial use, but it is not too soon to begin the effort.

## Additional Reading

A. K. Dewdney, "Nanotechnology," *Scientific American,* January 1988, 100–103.
K. E. Drexler, *Engines of Creation,* (Garden City, New York: Anchor Press/Doubleday, 1986).

K. E. Drexler, "Exploring Future Technologies," *Doing Science: The Reality Club*, ed. J. Brockman (New York: Prentice Hall, 1988).

J. Lehn, "Supramolecular Chemistry," *Angewandte Chemie, International English Edition*, 27 (1988): 90–112.

S. Vogel, "The Shape of Proteins to Come," *Discover*, October 1988, 38–43.

# Appendix B: There's Plenty of Room at the Bottom: An Invitation to Enter a New Field of Physics

Richard Feynman

I imagine experimental physicists must often look with envy at men like Kamerlingh Onnes who discovered a field like low temperature that seems to be bottomless and in which one can go down and down. Such a man is then a leader and has some temporary monopoly in a scientific adventure. Percy Bridgman, in designing a way to obtain higher pressures, opened up another new field and was able to move into it and to lead us all along. The development of ever higher vacuum was a continuing development of the same kind.

I would like to describe a field, in which little has been done, but in which an enormous amount can be done in principle. This field is not quite the same as the others in that it will not tell us much of fundamental physics (in the sense of, "What are the strange particles?"), but it is more like solid-state physics in the sense that it might tell us much of great interest about the strange phenomena that occur in complex situations. Furthermore, a point that is most important is that it would have an enormous number of technical applications.

What I want to talk about is the problem of manipulating and controlling things on a small scale.

As soon as I mention this, people tell me about miniaturization, and how far it has progressed today. They tell me about electric motors that are the size of the nail on your small finger. And there is a device on the market, they tell me, by which you can write the Lord's Prayer on the head of a pin. But that's nothing; that's the most primitive, halting step

Reprinted with permission from *Engineering and Science* magazine, February 1960, © California Institute of Technology.

in the direction I intend to discuss. It is a staggeringly small world that is below. In the year 2000, when they look back at this age, they will wonder why it was not until the year 1960 that anybody began seriously to move in this direction.

*Why cannot we write the entire 24 volumes of the* Encyclopaedia Britannica *on the head of a pin?*

Let's see what would be involved. The head of a pin is a sixteenth of an inch across. If you magnify it by 25,000 diameters, the area of the head of the pin is then equal to the area of all the pages of the *Encyclopaedia Britannica*. Therefore, all it is necessary to do is to reduce in size all the writing in the Encyclopaedia by 25,000 times. Is that possible? The resolving power of the eye is about 1/120 of an inch—that is roughly the diameter of one of the little dots on the fine half-tone reproductions in the Encyclopaedia. This, when you demagnify it by 25,000 times, is still 80 angstroms in diameter—32 atoms across, in an ordinary metal. In other words, one of those dots still would contain in its area 1,000 atoms. So, each dot can easily be adjusted in size as required by the photoengraving, and there is no question that there is enough room on the head of a pin to put all of the *Encyclopaedia Britannica*.

Furthermore, it can be read if it is so written. Let's imagine that it is written in raised letters of metal; that is, where the black is in the Encyclopedia, we have raised letters of metal that are actually 1/25,000 of their ordinary size. How would we read it?

If we had something written in such a way, we could read it using techniques in common use today. (They will undoubtedly find a better way when we do actually have it written, but to make my point conservatively I shall just take techniques we know today.) We would press the metal into a plastic material and make a mold of it, then peel the plastic off very carefully, evaporate silica into the plastic to get a very thin film, then shadow it by evaporating gold at an angle against the silica so that all the little letters will appear clearly, dissolve the plastic away from the silica film, and then look through it with an electron microscope!

There is no question that if the thing were reduced by 25,000 times in the form of raised letters on the pin, it would be easy for us to read it today. Furthermore; there is no question that we would find it easy to make copies of the master; we would just need to press the same metal plate again into plastic and we would have another copy.

## How Do We Write Small?

The next question is: How do we *write* it? We have no standard technique to do this now. But let me argue that it is not as difficult as it first appears to be. We can reverse the lenses of the electron microscope in order to demagnify as well as magnify. A source of ions, sent through the microscope lenses in reverse, could be focused to a very small spot. We could write with that spot like we write in a TV cathode ray oscilloscope, by going across in lines, and having an adjustment which determines the amount of material which is going to be deposited as we scan in lines.

This method might be very slow because of space charge limitations. There will be more rapid methods. We could first make, perhaps by some photo process, a screen which has holes in it in the form of the letters. Then we would strike an arc behind the holes and draw metallic ions through the holes; then we could again use our system of lenses and make a small image in the form of ions, which would deposit the metal on the pin.

A simpler way might be this (though I am not sure it would work): we take light and, through an optical microscope running backwards, we focus it onto a very small photoelectric screen. Then electrons come away from the screen where the light is shining. These electrons are focused down in size by the electron microscope lenses to impinge directly upon the surface of the metal. Will such a beam etch away the metal if it is run long enough? I don't know. If it doesn't work for a metal surface, it must be possible to find some surface with which to coat the original pin so that, where the electrons bombard, a change is made which we could recognize later.

There is no intensity problem in these devices—not what you are used to in magnification, where you have to take a few electrons and spread them over a bigger and bigger screen; it is just the opposite. The light which we get from a page is concentrated onto a very small area so it is very intense. The few electrons which come from the photoelectric screen are demagnified down to a very tiny area so that, again, they are very intense. I don't know why this hasn't been done yet!

That's the *Encyclopaedia Britannica* on the head of a pin, but let's consider all the books in the world. The Library of Congress has approximately 9 million volumes; the British Museum Library has 5 million volumes; there are also 5 million volumes in the National Library in France. Undoubtedly there are duplications, so let us say that there are some 24 million volumes of interest in the world.

What would happen if I print all this down at the scale we have been discussing? How much space would it take? It would take, of course, the area of about a million pinheads because, instead of there being just the 24 volumes of the Encyclopaedia, there are 24 million volumes. The million pinheads can be put in a square of a thousand pins on a side, or an area of about 3 square yards. That is to say, the silica replica with the paper-thin backing of plastic, with which we have made the copies, with all this information, is on an area of approximately the size of 35 pages of the Encyclopaedia. That is about half as many pages as there are in this magazine [*Engineering and Science*]. All of the information which all of mankind has ever recorded in books can be carried around in a pamphlet in your hand—and not written in code, but a simple reproduction of the original pictures, engravings, and everything else on a small scale without loss of resolution.

What would our librarian at Caltech say, as she runs all over from one building to another, if I tell her that, ten years from now, all of the information that she is struggling to keep track of—120,000 volumes, stacked from the floor to the ceiling, drawers full of cards, storage rooms full of the older books—can be kept on just one library card! When the University of Brazil, for example, finds that their library is burned, we can send them a copy of every book in our library by striking off a copy from the master plate in a few hours and mailing it in an envelope no bigger or heavier than any other ordinary airmail letter.

Now, the name of this talk is "There is *Plenty* of Room at the Bottom"—not just "There is Room at the Bottom." What I have demonstrated is that there *is* room—that you can decrease the size of things in a practical way. I now want to show that there is *plenty* of room. I will not now discuss how we are going to do it, but only what is possible in principle—in other words, what is possible according to the laws of physics. I am not inventing antigravity, which is possible someday only if the laws are not what we think. I am telling you what could be done if the laws are what we think; we are not doing it simply because we haven't yet gotten around to it.

## Information on a Small Scale

Suppose that, instead of trying to reproduce the pictures and all the information directly in its present form, we write only the information content in a code of dots and dashes, or something like that, to represent the various letters. Each letter represents six or seven "bits" of information; that is, you need only about six or seven dots or dashes for each letter. Now, instead of writing everything, as I did before, on the surface of the head of a pin, I am going to use the interior of the material as well.

Let us represent a dot by a small spot of one metal, the next dash by an adjacent spot of another metal, and so on. Suppose, to be conservative, that a bit of information is going to require a little cube of atoms $5 \times 5 \times 5$—that is 125 atoms. Perhaps we need a hundred and some odd atoms to make sure that the information is not lost through diffusion, or through some other process.

I have estimated how many letters there are in the encyclopaedia, and I have assumed that each of my 24 million books is as big as an encyclopaedia volume, and have calculated, then, how many bits of information there are ($10^{15}$). For each bit I allow 100 atoms. And it turns out that all of the information that man has carefully accumulated in all the books in the world can be written in this form in a cube of material one two-hundredth of an inch wide—which is the barest piece of dust that can be made out by the human eye. So there is *plenty* of room at the bottom! Don't tell me about microfilm!

This fact—that enormous amounts of information can be carried in an exceedingly small space—is, of course, well known to the biologists, and resolves the mystery which existed before we understood all this clearly, of how it could be that, in the tiniest cell, all of the information for the organization of a complex creature such as ourselves can be stored. All this information—whether we have brown eyes, or whether we think at all, or that in the embryo the jawbone should first develop with a little hole in the side so that later a nerve can grow through it—all this information is contained in a very tiny fraction of the cell in the form of long-chain DNA molecules in which approximately 50 atoms are used for one bit of information about the cell.

## Better Electron Microscopes

If I have written in a code, with $5 \times 5 \times 5$ atoms to a bit, the question is: How could I read it today? The electron microscope is not quite good enough; with the greatest care and effort, it can only resolve about 10 angstroms. I would like to try and impress upon you, while I am talking about all of these things on a small scale, the importance of improving the electron microscope by a hundred times. It is not impossible; it is not against the laws of diffraction of the electron. The wavelength of the electron in such a microscope is only 1/20 of an angstrom. So it should be possible to see the individual atoms. What good would it be to see individual atoms distinctly?

We have friends in other fields—in biology, for instance. We physicists often look at them and say, "You know the reason you fellows are making so little progress?" (Actually I don't know any field where they are making more rapid progress than they are in biology today.) "You should use more mathematics, like we do." They could answer us—but they're polite, so I'll answer for them: "What *you* should do in order for *us* to make more rapid progress is to make the electron microscope 100 times better."

What are the most central and fundamental problems of biology today? They are questions like: What is the sequence of bases in the DNA? What happens when you have a mutation? How is the base order in the DNA connected to the order of amino acids in the protein? What is the structure of the RNA; is it single-chain or double-chain, and how is it related in its order of bases to the DNA? What is the organization of the microsomes? How are proteins synthesized? Where does the RNA go? How does it sit? Where do the proteins sit? Where do the amino acids go in? In photosynthesis, where is the chlorophyll; how is it arranged; where are the carotenoids involved in this thing? What is the system of the conversion of light into chemical energy?

It is very easy to answer many of these fundamental biological questions; you just *look at the thing!* You will see the order of bases in the chain; you will see the structure of the microsome. Unfortunately, the present microscope sees at a scale which is just a bit too crude. Make the microscope one hundred times more powerful, and many problems of biology would be made very much easier. I exaggerate, of course, but the biologists would surely be very thankful to you—and they would prefer that to the criticism that they should use more mathematics.

The theory of chemical processes today is based on theoretical physics. In this sense, physics supplies the foundation of chemistry. But chemistry also has analysis. If you have a strange substance and you want to know what it is, you go through a long and complicated process of chemical analysis. You can analyze almost anything today, so I am a little late with my idea. But if the physicists wanted to, they could also dig under the chemists in the problem of chemical analysis. It would be very easy to make an analysis of any complicated chemical substance; all one would have to do would be to look at it and see where the atoms are. The only trouble is that the electron microscope is one hundred times too poor. (Later, I would like to ask the question: Can the physicists do something about the third problem of chemistry—namely, synthesis? Is there a *physical* way to synthesize any chemical substance?)

The reason the electron microscope is so poor is that the *f*-value of the lenses is only 1 part to 1000; you don't have a big enough numerical aperture. And I know that there are theorems which prove that it is impossible, with axially symmetrical stationary field lenses, to produce an *f*-value any bigger than so and so; and therefore the resolving power at the present time is at its theoretical maximum. But in every theorem there are assumptions. Why must the field be axially symmetrical? Why must the field be stationary? Can't we have pulsed electron beams in fields moving up along with the electrons? Must the field be symmetrical? I put this out as a challenge: Is there no way to make electron microscopes more powerful?

### The Marvelous Biological System

The biological example of writing information on a small scale has inspired me to think of something that should be possible. Biology is not simply writing information; it is *doing something* about it. A biological system can be exceedingly small. Many of the cells are very tiny, but they are very active; they manufacture various substances; they walk around; they wiggle; and they do all kinds of marvelous things—all on a very small scale. Also, they store information. Consider the possibility that we too can make a thing very small which does what we want—that we can manufacture an object that maneuvers at that level!

There may even be an economic point to this business of making things very small. Let me remind you of some of the problems of computing machines. In computers we have to store an enormous amount of information. The kind of writing that I was mentioning before, in which I had everything down as a distribution of metal, is permanent. Much more interesting to a computer is a way of writing, erasing, and writing something else. (This is usually because we don't want to waste the material on which we have just written. Yet if we could write it in a very small space, it wouldn't make any difference; it could just be thrown away after it was read. It doesn't cost very much for the material.)

### Miniaturizing the Computer

I don't know how to do this on a small scale in a practical way, but I do know that computing machines are very large; they fill rooms. Why can't we make them very small, make them of little wires, little elements—and by little, I mean *little*. For instance, the wires should be 10 or 100 atoms in diameter, and the circuits should be a few thousand angstroms across. Everybody who has analyzed the logical theory of computers has come to the conclusion that the possibilities of computers are very interesting—if they could be made to be more complicated by several orders of magnitude. If they had millions of times as many elements, they could make judgments. They would have time to calculate what is the best way to make the calculation that they are about to make. They could select the method of analysis which, from their experience, is better than the one that we would give to them. And in many other ways, they would have new qualitative features.

If I look at your face I immediately recognize that I have seen it before. (Actually, my friends will say I have chosen an unfortunate example here for the subject of this illustration. At least I recognize that it is a *man* and not an *apple*.) Yet there is no machine which, with that speed, can take a picture of a face and say even that it is a man; and much less that it is the same man that you showed it before—unless it is exactly the same picture. If the face is changed; if I am closer to the face; if I am further from the face; if the light changes—I recognize it anyway. Now, this little computer I carry in my head is easily able to do that. The computers that we build are not able to do that. The number of elements in this bone box of mine are enormously greater than the number of elements in our "wonderful" computers. But our mechanical computers are too big; the elements in this box are microscopic. I want to make some that are *sub*-microscopic.

If we wanted to make a computer that had all these marvelous extra qualitative abilities, we would have to make it, perhaps, the size of the Pentagon. This has several disadvantages. First, it requires too much material; there may not be enough germanium in the world for all the transistors which would have to be put into this enormous thing. There is also the problem of heat generation and power consumption; TVA would be needed to run the computer. But an even more practical difficulty is that the computer would be limited to a certain speed. Because of its large size, there is finite time required to get the information from one place to another. The information cannot go any faster than the speed of light—so, ultimately, when our computers get faster and faster and more and more elaborate, we will have to make them smaller and smaller.

But there is plenty of room to make them smaller. There is nothing that I can see in the physical laws that says the computer elements cannot be made enormously smaller than they are now. In fact, there may be certain advantages.

**Miniaturization by Evaporation**

How can we make such a device? What kind of manufacturing processes would we use? One possibility we might consider, since we have talked about writing by putting atoms down in a certain arrangement, would be to evaporate the material, then evaporate the insulator next to it.

Then, for the next layer, evaporate another position of a wire, another insulator, and so on. So, you simply evaporate until you have a block of stuff which has the elements—coils and condensers, transistors and so on—of exceedingly fine dimensions.

But I would like to discuss, just for amusement, that there are other possibilities. Why can't we manufacture these small computers somewhat like we manufacture the big ones? Why can't we drill holes, cut things, solder things, stamp things out, mold different shapes all at an infinitesimal level? What are the limitations as to how small a thing has to be before you can no longer mold it? How many times when you are working on something frustratingly tiny like your wife's wrist watch, have you said to yourself, "If I could only train an ant to do this!" What I would like to suggest is the possibility of training an ant to train a mite to do this. What are the possibilities of small but movable machines? They may or may not be useful, but they surely would be fun to make.

Consider any machine—for example, an automobile—and ask about the problems of making an infinitesimal machine like it. Suppose, in the particular design of the automobile, we need a certain precision of the parts; we need an accuracy, let's suppose, of 4/10,000 of an inch. If things are more inaccurate than that in the shape of the cylinder and so on, it isn't going to work very well. If I make the thing too small, I have to worry about the size of the atoms; I can't make a circle out of "balls" so to speak, if the circle is too small. So, if I make the error, corresponding to 4/10,000 of an inch, correspond to an error of 10 atoms, it turns out that I can reduce the dimensions of an automobile 4,000 times, approximately—so that it is 1 mm across. Obviously, if you redesign the car so that it would work with a much larger tolerance, which is not at all impossible, then you could make a much smaller device.

It is interesting to consider what the problems are in such small machines. Firstly, with parts stressed to the same degree, the forces go as the area you are reducing, so that things like weight and inertia are of relatively no importance. The strength of material, in other words, is very much greater in proportion. The stresses and expansion of the flywheel from centrifugal force, for example, would be the same proportion only if the rotational speed is increased in the same proportion as we decrease the size. On the other hand, the metals that we use have a grain structure, and this would be very annoying at small scale because the material is not homogeneous. Plastics and glass and things of this amorphous nature are very much more homogeneous, and so we would have to make our machines out of such materials.

There are problems associated with the electrical part of the system—with the copper wires and the magnetic parts. The magnetic properties on a very small scale are not the same as on a large scale; there is the "domain" problem involved. A big magnet made of millions of domains can only be made on a small scale with one domain. The electrical equipment won't simply be scaled down; it has to be redesigned. But I can see no reason why it can't be redesigned to work again.

### Problems of Lubrication

Lubrication involves some interesting points. The effective viscosity of oil would be higher and higher in proportion as we went down (and if we increase the speed as much as we can). If we don't increase the speed so much, and change from oil to kerosene or some other fluid, the problem is not so bad. But actually we may not have to lubricate at all! We have a lot of extra force. Let the bearings run dry; they won't run hot because the heat escapes away from such a small device very, very rapidly.

This rapid heat loss would prevent the gasoline from exploding, so an internal combustion engine is impossible. Other chemical reactions, liberating energy when cold, can be used. Probably an external supply of electrical power would be most convenient for such small machines.

What would be the utility of such machines? Who knows? Of course, a small automobile would only be useful for the mites to drive around in, and I suppose our Christian interests don't go that far. However, we did note the possibility of the manufacture of small elements for computers in completely automatic factories, containing lathes and other machine tools at the very small level. The small lathe would not have to be exactly like our big lathe. I leave to your imagination the improvement of the design to take full advantage of the properties of things on a small scale, and in such a way that the fully automatic aspect would be easiest to manage.

A friend of mine (Albert R. Hibbs) suggests a very interesting possibility for relatively small machines. He says that, although it is a very wild idea, it would be interesting in surgery if you could swallow the surgeon. You put the mechanical surgeon inside the blood vessel and it goes into the heart and "looks" around. (Of course the information has to be fed out.) It finds out which valve is the faulty one and takes a little knife and slices it out. Other small machines might be permanently incorporated in the body to assist some inadequately functioning organ.

Now comes the interesting question: How do we make such a tiny mechanism? I leave that to you. However, let me suggest one weird possibility. You know, in the atomic energy plants they have materials and machines that they can't handle directly because they have become radioactive. To unscrew nuts and put on bolts and so on, they have a set of master and slave hands, so that by operating a set of levers here, you control the "hands" there, and can turn them this way and that so you can handle things quite nicely.

Most of these devices are actually made rather simply, in that there is a particular cable, like a marionette string, that goes directly from the controls to the "hands." But, of course, things also have been made using servo motors, so that the connection between the one thing and the other is electrical rather than mechanical. When you turn the levers, they turn a servo motor, and it changes the electrical currents in the wires, which repositions a motor at the other end.

Now, I want to build much the same device—a master-slave system which operates electrically. But I want the slaves to be made especially carefully by modern large-scale machinists so that they are one-fourth the scale of the "hands" that you ordinarily maneuver. So you have a scheme by which you can do things at one-quarter scale anyway—the little servo motors with little hands play with little nuts and bolts; they drill little holes; they are four times smaller. Aha! So I manufacture a quarter-size lathe; I manufacture quarter-size tools; and I make, at the one-quarter scale, still another set of hands again relatively one-quarter size! This is one-sixteenth size, from my point of view. And after I finish doing this I wire directly from my large-scale system, through transformers perhaps, to the one-sixteenth-size servo motors. Thus I can now manipulate the one-sixteenth-size hands.

Well, you get the principle from there on. It is rather a difficult program, but it is a possibility. You might say that one can go much farther in one step than from one to four. Of course, this has all to be designed very carefully and it is not necessary simply to make it like hands. If you thought of it very carefully, you could probably arrive at a much better system for doing such things.

If you work through a pantograph, even today, you can get much more than a factor of four in even one step. But you can't work directly through a pantograph which makes a smaller pantograph which then makes a smaller pantograph—because of the looseness of the holes and the irregularities of construction. The end of the pantograph wiggles with a relatively greater irregularity than the irregularity with which you move your hands. In going down this scale, I would find the end of the pantograph on the end of the pantograph on the end of the pantograph shaking so badly that it wasn't doing anything sensible at all.

At each stage, it is necessary to improve the precision of the apparatus. If, for instance, having made a small lathe with a pantograph, we find its lead screw irregular—more irregular than the large-scale one—we could lap the lead screw against breakable nuts that you can reverse in the usual way back and forth until this lead screw is, at its scale, as accurate as our original lead screws, at our scale.

We can make flats by rubbing unflat surfaces in triplicates together— in three pairs—and the flats then become flatter than the thing you started with. Thus, it is not impossible to improve precision on a small scale by the correct operations. So, when we build this stuff, it is necessary at each step to improve the accuracy of the equipment by working for awhile down there, making accurate lead screws, Johansen blocks, and all the other materials which we use in accurate machine work at the higher level. We have to stop at each level and manufacture all the stuff to go to the next level—a very long and very difficult program. Perhaps you can figure a better way than that to get down to small scale more rapidly.

Yet, after all this, you have just got one little baby lathe four thousand times smaller than usual. But we were thinking of making an enormous computer, which we were going to build by drilling holes on this lathe to make little washers for the computer. How many washers can you manufacture on this one lathe?

## A Hundred Tiny Hands

When I make my first set of slave "hands" at one-fourth scale, I am going to make ten sets. I make ten sets of "hands," and I wire them to my original levers so they each do exactly the same thing at the same time in parallel. Now, when I am making my new devices one-quarter again as small, I let each one manufacture ten copies, so that I would have a hundred "hands" at the 1/16th size.

Where am I going to put the million lathes that I am going to have? Why, there is nothing to it; the volume is much less than that of even one full-scale lathe. For instance, if I made a billion little lathes, each 1/4,000 of the scale of a regular lathe, there are plenty of materials and space available because in the billion little ones there is less than two percent of the materials in one big lathe.

It doesn't cost anything for materials, you see. So I want to build a billion tiny factories, models of each other, which are manufacturing simultaneously, drilling holes, stamping parts, and so on.

As we go down in size, there are a number of interesting problems that arise. All things do not simply scale down in proportion. There is the problem that materials stick together by the molecular (Van der Waals) attractions. It would be like this: After you have made a part and you unscrew the nut from a bolt, it isn't going to fall down because the gravity isn't appreciable; it would even be hard to get it off the bolt. It would be like those old movies of a man with his hands full of molasses, trying to get rid of a glass of water. There will be several problems of this nature that we will have to be ready to design for.

## Rearranging the Atoms

But I am not afraid to consider the final question as to whether, ultimately—in the great future—we can arrange the atoms the way we want; the very *atoms,* all the way down! What would happen if we could arrange the atoms one by one the way we want them (within reason, of course; you can't put them so that they are chemically unstable, for example).

Up to now, we have been content to dig in the ground to find minerals. We heat them and we do things on a large scale with them, and we hope to get a pure substance with just so much impurity, and so on. But we must always accept some atomic arrangement that nature gives us. We haven't got anything, say, with a "checkerboard" arrangement, with the impurity atoms exactly arranged 1000 angstroms apart, or in some other particular pattern.

What could we do with layered structures with just the right layers? What would the properties of materials be if we could really arrange the atoms the way we want them? They would be very interesting to investigate theoretically. I can't see exactly what would happen, but I can hardly doubt that when we have some control of the arrangement of things on a small scale we will get an enormously greater range of possible properties that substances can have, and of different things that we can do.

Consider, for example, a piece of material in which we make little coils and condensers (or their solid state analogs) 1000 or 10,000 angstroms in a circuit, one right next to the other, over a large area, with little antennas sticking out at the other end—a whole series of circuits.

Is it possible, for example, to emit light from a whole set of antennas, like we emit radio waves from an organized set of antennas to beam the radio programs to Europe? The same thing would be to *beam* the light out in a definite direction with very high intensity. (Perhaps such a beam is not very useful technically or economically.)

I have thought about some of the problems of building electric circuits on a small scale, and the problem of resistance is serious. If you build a corresponding circuit on a small scale, its natural frequency goes up, since the wave length goes down as the scale; but the skin depth only decreases with the square root of the scale ratio, and so resistive problems are of increasing difficulty. Possibly we can beat resistance through the use of superconductivity if the frequency is not too high, or by other tricks.

### Atoms in a Small World

When we get to the very, very small world—say circuits of seven atoms— we have a lot of new things that would happen that represent completely new opportunities for design. Atoms on a small scale behave like *nothing* on a large scale, for they satisfy the laws of quantum mechanics. So, as we go down and fiddle around with the atoms down there, we are working with different laws, and we can expect to do different things. We can manufacture in different ways. We can use, not just circuits, but some system involving the quantized energy levels, or the interactions of quantized spins, etc.

Another thing we will notice is that, if we go down far enough, all of our devices can be mass produced so that they are absolutely perfect copies of one another. We cannot build two large machines so that the dimensions are exactly the same. But if your machine is only 100 atoms high, you only have to get it correct to one-half of one percent to make sure the other machine is exactly the same size—namely, 100 atoms high!

At the atomic level, we have new kinds of forces and new kinds of possibilities, new kinds of effects. The problems of manufacture and reproduction of materials will be quite different. I am, as I said, inspired by the biological phenomena in which chemical forces are used in repetitious fashion to produce all kinds of weird effects (one of which is the author).

The principles of physics, as far as I can see, do not speak against the possibility of maneuvering things atom by atom. It is not an attempt to violate any laws; it is something, in principle, that can be done; but in practice, it has not been done because we are too big.

Ultimately, we can do chemical synthesis. A chemist comes to us and says, "Look, I want a molecule that has the atoms arranged thus and so; make me that molecule." The chemist does a mysterious thing when he wants to make a molecule. He sees that it has got that ring, so he mixes this and that, and he shakes it, and he fiddles around. And, at the end of a difficult process, he usually does succeed in synthesizing what he wants. By the time I get my devices working, so that we can do it by physics, he will have figured out how to synthesize absolutely anything, so that this will really be useless.

But it is interesting that it would be, in principle, possible (I think) for a physicist to synthesize any chemical substance that the chemist writes down. Give the orders and the physicist synthesizes it. How? Put the atoms down where the chemist says, and so you make the substance. The problems of chemistry and biology can be greatly helped if our ability to see what we are doing, and to do things on an atomic level, is ultimately developed—a development which I think cannot be avoided.

Now, you might say, "Who should do this and why should they do it?" Well, I pointed out a few of the economic applications, but I know that the reason that you would do it might be just for fun. But have some fun! Let's have a competition between laboratories. Let one laboratory make a tiny motor which it sends to another lab which sends it back with a thing that fits inside the shaft of the first motor.

**High School Competition**

Just for the fun of it, and in order to get kids interested in this field, I would propose that someone who has some contact with the high schools think of making some kind of high school competition. After all, we haven't even started in this field, and even the kids can write smaller than has ever been written before. They could have competition in high schools. The Los Angeles high school could send a pin to the Venice high school on which it says, "How's this?" They get the pin back, and in the dot of the "i" it says, "Not so hot."

Perhaps this doesn't excite you to do it, and only economics will do so. Then I want to do something; but I can't do it at the present moment, because I haven't prepared the ground. It is my intention to offer a prize of $1,000 to the first guy who can take the information on the page of a book and put it on an area 1/25,000 smaller in linear scale in such manner that it can be read by an electron microscope.

And I want to offer another prize—if I can figure out how to phrase it so that I don't get into a mess of arguments about definitions—of another $1,000 to the first guy who makes an operating electric motor which can be controlled from the outside and, not counting the lead-in wires, is only 1/64 inch cube.

I do not expect that such prizes will have to wait very long for claimants.[1]

**Notes**

1. William McLellan claimed the prize for a 1/64th-inch electric motor a few months after Feynman's talk. In 1985 Tom Newman was the first to reduce a book page to 1/25,000th its original size.

# Contributors

**Robert R. Birge**   received his B.S. in chemistry from Yale University in 1968 and his Ph.D. in chemistry from Wesleyan University in 1972. He was an NIH postdoctoral fellow in chemistry at Harvard University from 1973 to 1975. From 1975 to 1984 Dr. Birge was an assistant and then associate professor of chemistry at the University of California, Riverside. He then moved to Carnegie-Mellon University where he was Professor and Head of the Department of Chemistry. Since 1988 Dr. Birge has been the Director of the Center for Molecular Electronics at Syracuse University as well as a professor of Chemistry. The author or coauthor of over a hundred papers, Dr. Birge is a contributor to and editor for the British journal *Nanotechnology*.

**Federico Capasso**   received his doctorate in physics from the University of Rome, Italy, in 1973. He joined AT&T Bell Labs as a member of the technical staff in 1977. Since 1987 he has been head of the Quantum Phenomena and Device Research Department. He is a recipient of the 1991 IEEE David Sarnoff Award, the 1984 AT&T Distinguished Member of Technical Staff Award, and the 1984 Award of Excellence of the Society for Technical Communications. He is a fellow of the IEEE, the American Physical Society, the Optical Society of America, and the International Society for Optical Engineering. Dr. Capasso has coauthored over 150 papers and holds 20 U.S. Patents.

**BC Crandall**   is cofounder and vice president of Prime Arithmetics, Inc., a California corporation dedicated to the design and implementation of object-oriented, parallel-processing operating systems and computing architectures. He received his B.A. from St. Lawrence University in 1981 and has edited technical and scientific books and papers since 1985. He is the editor of *Nanotechnology and the Culture of Abundance* (MIT Press, 1993), a collection of papers on the technical potential of nanotechnology.

**Eric Drexler**   received a B.S. degree in interdisciplinary science, an M.S. degree in engineering (supported by a National Science Foundation Fellowship), and a Ph.D. in molecular nanotechnology, all from MIT. He is currently a visiting scholar at Stanford University's Department of Computer Science.

Eric Drexler is a researcher concerned with emerging technologies and their consequences for the future. This interest led him to perform ground-breaking studies in the field of molecular nanotechnology—an anticipated technology based on molecular machines able to build objects to complex atomic specifications. He has written a series of technical papers on nanotechnology and two books, *Engines of Creation* (Doubleday, 1986) and *Unbounding the Future* (with Chris Peterson and Gayle Pergamit; Morrow, 1991), that describe the prospects ahead and some strategies for dealing with them. To help cope with the opportunities and dangers presented by this new field, he founded the MIT Nanotechnology Study Group and now serves as president of the Foresight Institute, a nonprofit educational organization founded in 1986 to help prepare for advanced technologies. In spring 1988 he taught (at Stanford University) the first formal course on nanotechnology and exploratory engineering. He chaired the First Foresight Conference on Nanotechnology in October, 1989, as well as the second conference in November, 1991, sponsored by Stanford University's Department of Materials Science and Engineering, the University of Tokyo's Research Center for Advanced Science and Technology, and the Foresight Institute. He is completing a technical book on molecular nanotechnology to be published by John Wiley & Sons in 1992.

**Gregory M. Fahy**   received his B.S. in biology from the University of California, Irvine, in 1972, where he team-taught a for-credit course on life extension as a senior. He received his Ph.D. in pharmacology and cryobiology from the Medical College of Georgia in 1977, working on basic mechanisms of freezing injury as a pathway to whole-organ cryopreservation. His techniques for analyzing freezing injury have had a lasting influence on some of the best minds in the field. Dr. Fahy was a postdoctoral fellow at the American Red Cross from 1977 to 1980. Following this fellowship, he remained at the American Red Cross as a research associate (1980–1984), research scientist (1984), and scientist II (1984–present), where he has also been a project leader since 1987. In 1980, he conceived of the idea of vitrifying whole organs instead of freezing them, a goal he has made steady progress on ever since and which has revolutionized the field of cryobiology in several respects.

Dr. Fahy has received a number of institutional honors and has over seventy formal publications as well as many informal publications. In addition, he belongs to several professional societies including the American Aging Association, the Society for Cryobiology, and the Transplantation Society. Dr. Fahy, who has been invited to speak at over twenty seminars and symposia and has appeared more than a dozen times on radio and television, is actively pursing his interests in organ preservation, vitrification, baroinjury and baroprotection, chilling injury, cryoprotectant toxicity, and interventive gerontology.

**John S. Foster**   received a B.A. in physics from the University of California, San Diego, in 1980 and a Ph.D. in applied physics from Stanford University in 1984, where he also received IBM and ARCS fellowships. He joined the faculty at Stanford in 1984 in the Applied Physics Department as the Marvin Chodorow Fellow and in the same year received the IBM Faculty Development Award. His research at Stanford centered on developing an acoustic microscope with better than 200 Å resolution operating in superfluid helium, inventing a method for spatially imaging thermal phonon emissions from solids, and devising a quantum noise limited parametric amplifier for microwave frequencies. Foster joined the

Research Division at IBM, Almaden, in 1986 and in 1989 became the manager of Molecular Studies for Manufacturing in Storage Systems and Technology.

**Tracy Handel**   received her B.S. in chemistry from Bucknell University in 1980. She worked at Kodak for three years in a polymer physical chemistry group, then she attended the California Institute of Technology where she received her chemistry Ph.D. in 1989. Her thesis was on the biophysical properties of synthetic liposomes for use as drug delivery vehicles. Her postdoctoral work on protein design, under the direction of William F. DeGrado (currently at du Pont Merck Pharmaceuticals), focused on 4-helix bundle proteins. Dr. Handel recently joined du Pont Merck Pharmaceuticals as a principal investigator in the structural biology group where she currently works on protein design and the structural characterization of proteins using multidimensional nuclear magnetic resonance.

**Bill Joy**   one of the founders of Sun Microsystems, is the designer of the NFS Network File System and codesigner of the SPARC chip. Before cofounding Sun, Mr. Joy created the Berkeley version of the UNIX operating system, the standard for academic and scientific research in the late 1970s and early 1980s, and founded the BSD series of software distributions. He is the creator of the vi text editor. Mr. Joy has propounded "Joy's Law," which states that the performance of a personal microprocessor-based system should double each year, and predicts affordable 1000 MIPS desktop computers by 1995. He is currently directing Sun's research and development, to assure that Joy's Law becomes a reality, and exploring new computing paradigms for exploiting this new power.

**Arthur Kantrowitz**   has been professor of engineering at Thayer School, Dartmouth College since 1978. Previous to this position, he was the chairman and chief executive officer of the Avco Everett Research Laboratory, which he founded in 1955. His degrees are in physics from Columbia University. He has made scientific contributions to and pioneered work in physical gas dynamics, fluid mechanics, magnetohydrodynamic power generation, heart-assist devices, high-energy lasers, laser isotope separation, laser propulsion to earth orbit, and space reentry. He has also been concerned with the interactions of science and technology with society, particularly with the politicization of scientific controversy. Dr. Kantrowitz chaired the Presidential Task Force on the Science Court in 1976 to address this problem. He is a member of the National Academy of Sciences and the National Academy of Engineering.

**James B. Lewis**   is a scientist at Bristol-Myers Squibb Pharmaceutical Research Institute. His fields of study include virology, vaccine, and immunotherapy development. In 1972 he received a Ph.D. in chemistry from Harvard; his thesis was on the structure of RNA. After postdoctoral work in Switzerland, he worked as a senior scientist and faculty member at the Cold Spring Harbor Laboratory and at the Fred Hutchinson Cancer Research Center before joining Bristol-Myers Squibb in 1988. Dr. Lewis is author of more than forty scientific research publications in biochemistry, virology, and molecular biology and is a founding member of the Seattle nanotechnology study group.

**Joseph Mallon**   graduated cum laude from Farleigh Dickinson University with a B.S. in physics in 1976. From 1978 to 1985 he was vice president of engineering at Kulite Semiconductor Products, Inc. In 1985 he left Kulite to cofound NovaSensor, Inc., currently recognized as an international leader in micromachined pressure and acceleration sensors. Mallon has published approximately thirty papers on solid sensors and has been awarded nearly forty patents.

**Norman Margolus**   received a B.Sc. in physics from the University of Alberta where he was honored as the top physics student, in 1977, and his Ph.D. in physics from MIT in 1987. Since his graduation from MIT, Dr. Margolus has been a research scientist at the MIT Laboratory for Computer Science where he has been developing and constructing cellular automata architectures and machines. He has published several papers on physically ideal computation and physical modeling with cellular automata and has coauthored, with his collaborator Tommaso Toffoli, *Cellular Automata Machines: A New Environment for Modeling,* MIT Press, 1987. His next book, *Physics-Like Models of Computation,* MIT Press, is due out in 1993.

**Ralph Merkle**   received his Ph.D. in electrical engineering from Stanford University in 1979 where he coinvented public key cryptography. He pursued research in this area at Bell Northern Research until he joined Elxsi, San Jose, California, in 1981. He left Elxsi to join Xerox PARC in 1988, where he is pursuing research in computational nanotechnology. Dr. Merkle has five patents and has published extensively.

**Lester W. Milbrath**   is the director of the Research Program in Environment and Society and professor emeritus of political science as well as sociology at the State University of New York at Buffalo. From 1976 to 1987 he was director of the Environmental Studies Center at SUNY/Buffalo. From 1969 to 1976 he was associate provost for social science and director of the Social Science Research Institute. Prior to coming to SUNY/Buffalo in 1966, he served on the faculties of Northwestern University, Duke University, and the University of Tennessee. He holds a doctorate in political science from the University of North Carolina. He was a Fulbright Scholar to Norway in 1961–62 and again in 1972–73; in the fall term 1988, he was visiting Fulbright Professor at National Taiwan University in Taipei. He has also been a visiting professor at Aarhus University in Denmark, a visiting research scholar at the Center for Resource and Environmental Studies at the Australian National University in Canberra, and at the Institute for Research on Societal Developments at Mannheim University in Germany.

Professor Milbrath's research in political science has focused on lobbying, political participation, and political beliefs with a special interest in the relationship between the environment and society. He is the author of several books, including *The Washington Lobbyists,* 1963, *Environmentalists: Vanguard for a New Society,* 1984, and *Envisioning a Sustainable Society: Learning Our Way Out,* 1989. The latter volume summarizes a lifetime of learning about the predicament Western civilization has created for itself because of its attempt to dominate and distance itself from planet Earth's life systems.

**Hiroyuki Sasabe**   attended the University of Tokyo where he received B.S., M.S., and Ph.D. degrees in applied physics in 1964, 1966, and 1971, respectively. Dr. Sasabe worked as a research scientist in the Electrotechnical Laboratory of the

Ministry of International Trade and Industry (MITI) from 1966 to 1974. For the next nine years, he taught as an associate professor of electronic engineering at the Tokyo University of Agriculture and Technology. Since 1982, Dr. Sasabe has acted as the senior scientist and head of the Biopolymer Physics Laboratory, RIKEN (the Institute of Physical and Chemical Research). From 1986 to 1991 he was the coordinator of RIKEN's Frontier Research Program at the Bioelectronic Materials Laboratory and the Nonlinear Optics and Advanced Materials Laboratory. He has been team leader in the Laboratory for Nano-Photonics Materials since 1991. Dr. Sasabe has been a visiting professor at the Graduate School of Bio-Environmental Science at Saitama University since 1989. Dr. Sasabe has lectured at over a dozen universities and is a member of several professional societies including the Physical Society of Japan, the Institute of Electrical Engineering (Japan), the American Physical Society, and the Materials Research Society (United States).

**Gordon Tullock** is Karl Eller Professor of Economics and Political Science at the University of Arizona. Dr. Tullock is former president of the Southern Economic Association and the Public Choice Society, as well as the president-elect of the Western Economic Association and a member of the Board of the American Political Association. He is the author of a number of books including *The Calculus of Consent* (with James M. Buchanan), *Towards the Mathematics of Politics, Trials on Trial,* and *The Organization of Inquiry.*

**Michael D. Ward** received a B.A. in chemistry, summa cum laude, in 1977 from William Paterson College and a Ph.D. in chemistry from Princeton University in 1981. He then spent one year at the University of Texas at Austin as a Welch Postdoctoral Fellow. From 1982 to 1984 Dr. Ward served as project leader in the Fundamental Research Laboratories of the Standard Oil Company of Ohio. In 1984 he joined the staff of the Central Research and Development Department of E. I. duPont de Nemours, Inc. Dr. Ward joined the Department of Chemical Engineering and Materials Science at the University of Minnesota as an associate professor in 1990, where he is also a member of the graduate faculty of the Department of Chemistry.

Dr. Ward's research interests include design and fabrication of molecular materials, physical and electronic properties of molecular solids, investigation of crystallization mechanisms of molecular charge transfer solids, piezoelectric bio-sensors, and electrochemical methodologies, including the quartz crystal micro-balance and rotating ring-disk techniques. Dr. Ward has authored or coauthored over forty papers, holds four patents, and has been invited to speak at numerous institutions and scientific meetings.

# Index

Page numbers followed by an f indicate figures; n indicates the reference is a note.